FOSSILS ILLUSTRATED

VOLUME 1

GRAPTOLITES

WRITING IN THE ROCKS

GRAPTOLITES
WRITING IN THE ROCKS

EDITED BY
DOUGLAS PALMER
BARRIE RICKARDS

THE BOYDELL PRESS

First published 1991 by The Boydell Press, Woodbridge

The Boydell Press is an imprint of Boydell & Brewer Ltd
PO Box 9, Woodbridge, Suffolk IP12 3DF
and of Boydell & Brewer Inc.
PO Box 41026, Rochester, NY 14604, USA

ISBN 0 85115 262 7

British Library Cataloguing in Publication Data
Graptolites.
1. Graptolithina
1. Palmer, Douglas II. Rickards, Barrie III. Series
563.71
ISBN 0-85115-262-7

Library of Congress Cataloging-in-Publication Data
Graptolites : writing in the rocks / edited by Douglas Palmer, Barrie
Rickards.
p. cm. -- (Fossils illustrated ; v. 1)
Includes bibliographical references and index.
ISBN 0-85115-262-7 (alk. paper)
1. Graptolites. I. Palmer, Douglas. II. Rickards, Barrie.
III . Series.
QE840.5.G72 1990
563'.71--dc20 90-46784

FOSSILS ILLUSTRATED ISSN 0960-8664
Series editors: Douglas Palmer and Barrie Rickards

This publication is printed on acid-free paper

Printed in Great Britain by Whitstable Litho Printers Ltd, Whitstable, Kent

CONTENTS

FOREWORD

Fossils, the record of past life entombed in rocks of the earth's crust, have found many uses, ranging from exploration for natural resources to those as art objects, jewelry, and conversation pieces. Of those fossils that have found broad applicability in the quest for nature's resources and as fascinating curiosities as well, graptolites stand out. Graptolites are amongst the most fascinating of fossils because they are extinct and have left few clues to their ancestors or descendents (if, indeed, they had any) or even to what they looked like and how they lived.

Among extinct organisms, graptolites rank on a popularity scale at some significant distance from dinosaurs. Yet, graptolites have proven far more valuable from the perspectives of economic use and evolutionary patterns. To most collectors, graptolites appear to be mere bits of incongruous writing on rock. Indeed, most found are silhouettes. When three-dimensional, little-altered remains of graptolites' hard skeletal parts have been examined, however, these creatures have been found to have complex structures. The most commonly found graptolites seem to have lived a floating (planktic) life in the world's oceans. Certain of them lived as encrusting, mat-like features on firm marine bottoms. Others were small, conical or fan-shaped colonies that grew as shrubs upward from the sea floor. Shapes among colonies that floated changed dramatically through time. This feature has made graptolites a most useful fossil in documenting time synchroneities among widely-separated rock sequences as well as in proposing certain patterns in evolution. Most floating graptolites apparently preferred life in the warm tropical oceans of about 500 to 390 million years ago. Their patterns of occurrence suggest possible pathways of current flow in the ancient oceans in which they lived.

Graptolites are relatively easy to find in many parts of the world, especially in the British Isles. Because they are, this book's value lies in drawing attention to them. The reader is encouraged to find out about graptolites using the text and illustrations. As well, the reader is introduced to some of those learned scientists who, through their careers, wrestled with the problems of learning about an extinct group of fossils that all-too-often appear as little more than vague scribbles on rock. Debates about graptolites have enlivened the field of paleontology. A record of the controversies will entertain the reader. This compilation of knowledge about graptolites eloquently states the case for getting to know them.

Professor William B. N. Berry
Berkeley
California 94720
U.S.A.

ACKNOWLEDGEMENTS

We should especially like to thank Michael Howe for his early help with the project and Ken Harvey for printing up some photographs from a varied standard of negative and for doing the originals of a fair number of illustrations. Big G as a whole has been extremely helpful and we thank for their co-operation those members whose names do not actually appear in this book. We are pleased to record our appreciation of the help received from graptolithologists around the world. Those who are not Big G members and whose photographs have been used are credited in the photograph caption, but we thank all the following workers for photography and other assistance: Bill Berry (Berkeley, USA); Clive Burrett (Hobart, Australia); Berndt D. Erdtmann (West Berlin, Germany); Chen Xu (Nanjing, China); Charles Holland (Dublin); Hermann Jaeger (East Berlin, Germany); Nancy Kirk (Aberystwyth); Krystina Lindström (Lund, Sweden); Jörg Maletz (West Berlin, Germany); Mike Melchin (Waterloo, Canada); Florentin Paris (Rennes, France); Manfred Schauer (East Berlin, Germany); Peter Störch (Prague, Czechoslovakia); Lech Teller (Warsaw, Poland); Ni Yunan (Nanjing, China); Valdar Jaanusson (Sweden); and Henry Williams (St Johns, Newfoundland). The book was typed by Jacqui Hodkinson and Pat Hancock; and Gwynne Morris helped considerably, especially with photography. Richard Hughes, Adrian Rushton, Steve Tunnicliffe and Jan Zalasiewicz publish with the permission of the Director of the British Geological Survey, but the views expressed are necessarily their own and should not be taken to represent official BGS standpoints.

Dr Kirk is seen here, with Barrie Rickards,
at the 1985 meeting in Copenhagen of
the 3rd International Graptolite Conference

Big G dedicate this volume
to Nancy Hartshorne Kirk
in appreciation of her work on graptolites

INTRODUCTION

The ideas behind the writing of this book were several, but in large part it resulted from the British and Irish Graptolite Group (BIG G) members feeling that their chosen field of research was not too well represented in either popular books on palaeontology (or geology) or in textbooks. Discussions revealed that graptolite workers are commonly asked the very questions which we have used as chapter headings. That there is considerable interest in graptolites is always apparent in the protracted discussion/question period following public or academic lectures. The same discussions betray the fact that the considerable ignorance of people on this subject, academic and amateur alike, is a result of graptolithologists not presenting their field of study in anything like an attractive format, except perhaps at lectures. We can hardly expect someone with a latent or casual, intelligent, interest, to wade through a treatise or learned paper on graptolites. Furthermore, papers of this kind usually cover only a tiny proportion of the subject area, and very rarely have useful illustrations beyond the specific purpose of the paper.

In this tome we have attempted to illustrate the varied form of graptolites, for the most part exactly as they appear on the bedding plane at the time of collection. However, in order to emphasize that graptolite studies need not stop at the hand lens stage we have included a number of scanning electron microscope pictures, both at low magnifications and at high magnification. These relate to subjects discussed in the text, as do other illustrations. The taxonomic position of each graptolite illustrated is given as a number in brackets immediately after the scientific name: the number refers to Appendix 3. The illustrations are not arranged in a biologically systematic fashion, but rather in the order in which they are discussed in particular chapters and in relation to particular topics.

We do not claim that the text is an exhaustive coverage of graptolite research: rather, it reflects the areas in which we feel people have most interest. Equally, we have not photographed every single type of graptolite, but we have illustrated a fair diversity of forms – probably enough, in fact, to enable the enterprising teacher to illustrate the classification and evolution charts of Appendices 3 and 4! In general we have avoided text figures simply because, unlike photographs, they *are* available and readily usable in graptolite treatises.

We have been at pains to point out where debate exists, and that research does not always have cut and dried answers. In several areas of research the important questions have not yet been asked. There is, in graptolite research, considerable scope for enthusiastic amateur workers, at the very least by making important collections of good material.

On the contents page we have listed the main contributors to each chapter. However, it should be mentioned most members of BIG G have made a considerable input to the text as a whole, and minor additions to each chapter will have been made by others. The full list of contributors is given before the Index; and a full list of photographers who are not BIG G members is given in the acknowledgements. The scale of magnification is indicated on each photograph.

Terms are defined in the text at first usage. However, because a degree of independence is intended for each chapter, a glossary of terms is provided as Appendix 1. Again, because of the intended integrity some chapters may *briefly* duplicate what may have been dealt with in more detail elsewhere. A cross reference is given where appropriate. Because we are aware that you may have failed to find graptolites at localities where they were supposed to exist, we have provided, in Appendix 2, a list of collecting localities where you cannot possibly fail to succeed! Please note the words of caution at the beginning of Appendix 2, and the detailed instructions for finding the locality. Finally it may be that despite all our efforts in this book you may need further help in identifying what you have collected: to that end Appendix 5 lists those institutes that can help or, if they can't, they will know someone who can . . .

Douglas Palmer
Barrie Rickards

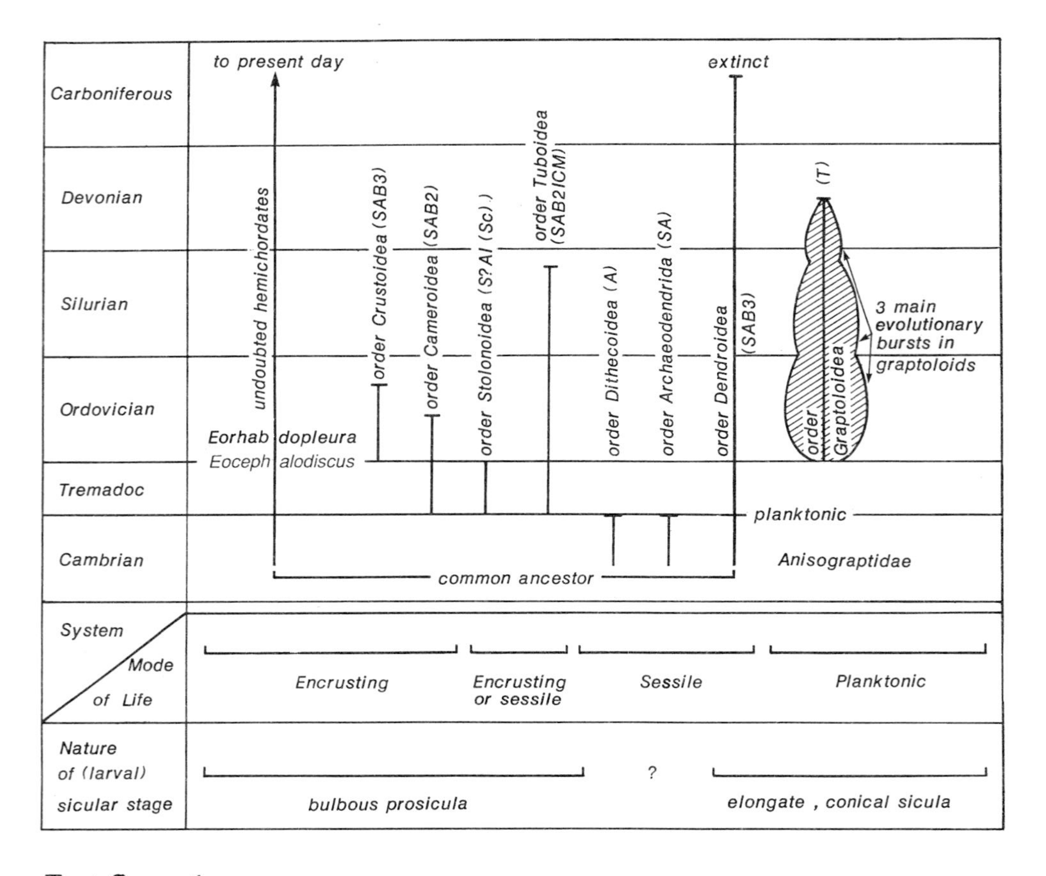

Text figure 1.

Overall classification of the phylum Hemichordate (i.e. graptolites and living hemichordata), with gross mode of life indicated and the nature of the larval phase. The evolution of the order Graptoloidea (shaded, to right) is given in more detail in Chapter 12 and Appendix 4; whilst the mode of life is discussed in Chapter 5. The very rare orders, Crustoidea to Archaeondendrida on this chart, are not illustrated in this book, but they are defined in terms of their component thecae and nature of stolonal budding (see Chapter 1). Thus on this diagram S = stolothecae; A = autothecae; B = bithecae; 3 = triad budding; 2 = diad budding; 1 = irregular stolons; C = conothecae; M = microthecae; 1(Sc) = irregular stolons which may be only partially sclerotised; T = thecae.

Chapter One

WHAT ARE THEY?
WHAT DO THEY LOOK LIKE?

Strange silvery-grey markings looking like pencil scribbles on rock surfaces gave rise to the name 'graptolith', from the Greek *graphein* = 'to write' and *lithos* = 'stone', now recognised as the fossil group known as graptolites. Strictly speaking *graptolithos* probably means scratching or incisions on the stone, but at least in a metaphorical sense they are unquestionably writings in the rocks. Many graptolites do indeed look at first sight to be no more than odd markings on a bedding plane in black shales or other rocks, but close inspection will often reveal them to be the completely flattened remains of graptolite periderm. These silvery, blackish, grey, or whitish flattened specimens may be composed of no more than thin films of carbon and/or chlorite, but surprisingly often show enough detail of the outlines to readily allow a full identification of the species. Indeed, a great deal of stratigraphic work is done based upon such graptolites.

Graptolites are *animals* despite their superficial resemblance to fern fronds or other small plants. Furthermore, they are *colonial* animals like corals and most bryozoans (moss animals) to which they are also similar in that the individual animals (zooids) of the colony were connected together with living tissue. Unlike these groups, however, the skeleton of the graptolite colony is usually much more geometric in its arrangement, and it is composed of a quite different substance. Whereas the skeleton in corals and bryozoans is made of largely calcareous, matter, that of the graptolites consists wholly of a tough, horny material, a fibrous protein called *collagen*, generally preserved as black carbon (Fig. 20). This material is commonly found today in vertebrate parts, such as bones, skin, and heart tissue.

Graptolites are extinct, ranging through time from the Middle Cambrian (550 million years ago) to the uppermost Carboniferous (300 million years ago) (Text fig. 1). This means that direct comparison with living creatures is very difficult; yet there is a consensus amongst the world's graptolite workers that the group is a class of hemichordate, related to living forms such as *Cephalodiscus* and *Rhabdopleura*, and so they are placed in the phylum Hemichordata (Figs 8, 9 and 33; and Appendix 3). The consensus need not be correct, of course, and a few researchers are of the opinion that graptolites are best regarded as an obscure group of coelenterates (the group containing corals) because, it is held, they had an external layer of soft living tissue (*extrathecal tissue*) connecting the zooids. The majority, dispute this and point to the great difficulty of distinguishing fossil rhabdopleurans in the Cambrian to Silurian periods, from graptolites occurring at the same time: they have skeletons of the same material, collagen, constructed in an almost identical fashion, and the zooids were similarly connected by a fine black rod called a *stolon* (Figs 33 and 34). Hence, in our Text fig. 1 we have taken the consensus view of basic relationships and basic evolution. These matters are discussed further in the following chapters (3, 12, Appendices 3 and 4) and it might be said that, like most matters scientific, the debate is under constant review!

We have already alluded to the fact that graptolites often look like writing in the rocks, or graphitic markings; that they can be obscure and also deceive. So how do we find and handle them? How do we recognise them? Actual collecting of specimens is dealt with in chapter 11; but it is useful, whilst collecting in the field, to know roughly what they can look like! Figs 26 – 30 depict some of the more common preservation modes. Some (Fig. 5) are indeed like rock hieroglyphs; others have an immediate aesthetic appeal because of the golden yellow pyrite, or Fool's Gold which infilled the colony and now shines through where the skeletal material (periderm) has flaked off (Fig. 25). Quite commonly, the fossilised colonies, called *rhabdosomes*, are silver or white on a blue-grey or grey-black mudrock, and as such are conspicuous (Figs 28 and 38). Pyritized specimens may be less conspicuous, whilst three-dimensional specimens which are neither infilled with pyrite, nor flattened to a bedding plane, but perhaps partly infilled by the mudrock itself, can be quite inconspicuous: such a preservation is shown in Figs 10 and 11 where the contrast between a fresh specimen and one treated carefully for photography, is also shown. Occasionally the preservation is peculiar in some way, such as alteration of pyrite to haematite, when the graptolite is red in colour; or to limonite, when it may be of yellow ochre colour. We have even seen them infilled by glauconite (green) and opaline silica (blue). In general, however, they are whitish (altered or infilled with a chloritic clay mineral material), dark coloured (carbonised periderm) or infilled with pyrite and therefore striking in appearance. Very commonly indeed the tiny tubes of the colony, called *thecae*, which housed the zooids during life, appear as tiny fret-saw blade like serrations along the branches (stipes) of the rhabdosomes (Figs 2 and 5). It cannot be emphasized too strongly that a hand lens is imperative for preliminary examination in the field: graptolites are only rarely microfossils; but the *detail* is not usually visible to the naked eye.

Fortunately there are certain lynch pins of reassuring fact which facilitate and support studies of graptolites. One is that all graptolite colonies grew from a small conical body called the *sicula*. This is labelled in Figs 17 and 23, and it can be seen in many subsequent illustrations in the book. The hollow sicula is remarkably uniform in shape, looking like a space rocket, elongate, narrow, conical (Fig. 13). However, there is a some variation that reflects the basic modes of life shown in Text fig. 1: those forms which live on the substratum, encrusting shells and stones tend, like today's rhabdopleurans, to have a boot-shaped sicula with a flat, encrusting portion; whereas those forms living as upright, bushy bottom dwellers have the apex of the sicular cone merging with a substrate such as a disc or fibrous bundle of 'roots'; however, the vast majority have the simple unmodified conical sicula of Figs 13–15.

The subject of sex in graptolites we treat fully in chapter 9. For the present suffice it to say that growth of the rhabdosome from the initial body or sicula was by budding, i.e. asexual development. The sicula itself was formed in two stages, an initial prosicula (like a pointed nose cone to a space rocket) and a later, added, metasicula (Figs 13 and 14). The prosicula has no growth stages, and the cone is always seen fully formed, though spiral strengthening rods and longitudinal rods may have been added (Fig. 13). The metasicula, in contrast was built by incremental small growth rings, or half rings with a zip-like or zig-zag contact (Fig. 15): the growth rings are called *fuselli*, and they typify all graptolites and many living hemichordates. The whole colony is built of these fuselli, often many thousands of them: their typical arrangement, is well seen on Figs 16 and 19. Thus the thecal tubes bud one from another (Fig. 22); and each tube itself it built up of numerous fuselli.

The fuselli can, therefore, conveniently be thought of as the building blocks of the colony. In addition to the fuselli there is a layer of plaster, usually on the outside of the fuselli, called *cortex*. This cortex is also composed entirely of fibrous collagen, but its substance is arranged differently to that of the fuselli: the latter has a spongy texture, the former comprises parallel bundles of *fibrils* (Fig. 20).

There is also a functional difference: whereas the fuselli are the building blocks, the cortex (of cortical layers) is probably a strengthening feature. The cortical layers actually consist of many criss-crossing, narrow bandage-like structures (Figs 15 and 19). The analogy with bandages has some substance, for in medical applications a criss-cross arrangement of bandage is often used for the strength it imparts to a structure. While both fuselli and cortex may be involved in the construction of features such as processes and spines, it is probably the cortex which gives strength to those structures (see also chapter 3).

In summary, then, all graptolites grew by asexual budding from an initial conical body, the sicula; they built the shape of both thecae and rhabdosome by increments of fuselli rings and half rings; and further shaped and strengthened the colony with layers of cortical bandages.

When the graptolites were bottom dwellers (*benthonic* forms) the cortical bandages were important in structuring the holdfast, and in strengthening any stem to the colony. However, one of the major debates in graptolite research today is exactly how 'root'-like structures could form; and how the thin thread-like structure growing from the apex of the sicula in planktonic forms, could be secreted (termed a *nema* in Fig. 18).

Peter Crowther and Barrie Rickards (1977) and Rickards and Lori Dumican (1984) have argued convincingly for a secretory mode like that of modern rhabdopleurans (Noel Dilly, 1986). But it seems unreasonable that such zooids could extend the 'root' system of a benthonic form by burying into the mud and adding increments to the 'root' tips! There are almost equally difficult problems when it comes to zooids secreting the tips of long *nemata* (pl. of nema) (Figs 106 and 107). For these reasons Nancy Kirk and Dennis Bates (1986) prefer that an extrathecal secretory membrane should have covered large parts of the colony and have been responsible for such features as as holdfasts and nemata (and derivatives of the latter such as three-vaned structures, Figs 24 and 38). The weakness of this particular argument is that it cannot explain as readily as the rhabdopleuran model the formation of bandages. The matter is, as yet, unresolved.

Thus whilst we can illustrate and describe various holdfasts, nemata, webs and spines (Figs 24 and 26 for example) we cannot yet fully explain their secretion. Time for secretion is also an unexplored avenue of research. We know from Noel Dilly's (1986) work that modern rhabdopleurans can take eight hours to secrete a single growth ring. They use the secretory cells of a disc (pre-oral lobe) in front of the mouth which becomes very soft and malleable, adjusting itself to the shape required, staying clamped in position for eight hours, and then lifting off to reveal a rapidly-hardening ring of spongy collagen. It is tempting to suppose that graptolites also did this, especially as the skeletal product is physically and biochemically the same, and no other groups use this system of skeletal construction. It is also now known that living rhabdopleurans and cephalodiscans secrete an outer, cortical layer. In the case of the encrusting rhabdopleurans this is sparse and thin, exactly as it is in the case of encrusting graptolites (e.g. camaroids and crustoids, Text fig. 1, or tuboids, Figs 45 and 46).

One further area of graptolite morphology needs further explanation. The *graptoloids* (Fig. 1) are relatively simple graptolites exhibiting the features described above. But the statement that each individual zooid was connected by soft, living tissue, is a supposition, a reasonable deduction, not a proven fact. No zooids or soft parts have been found in such groups and thecal tubes and stipe structure (Fig. 12) is quite simple, with no black stolon connecting successive thecae. At this point we shall curtail detailed discussions on such questions in relation to the Graptoloidea (Text fig. 1; Appendix 3), for they are examined in later chapters, but it is clear that providing the graptolites are at least moderately well-preserved then they are easily recognisable as graptolites by their characteristic preservations, saw-tooth appearance, and by their striking geometrical patterns (e.g. Figs 1, 2, 5 and 6). And this chapter has given us the opportunity to touch briefly on two areas of substantial debate, namely the affinities of graptolites, and the secretion of parts of the skeleton.

The connection of successive thecae by a black stolon is, in fact, a feature of most of the other orders. In these orders there is more than one thecal type (a formula defining each order in terms of its thecae is given in the explanation to Text fig. 1). There are two *main* kinds of thecae, called *autothecae* (homologous with the *thecae* of graptoloids) and *bithecae* (Fig. 16) which are usually much smaller and rather inconspicuous. The dendroid graptolites have a quite complicated internal structure revolving around the origin and early growth of these different thecal tubes within the stipe. The development may be summarised as follows.

Each black stolon (Fig. 34) thickens at regular intervals into a node and immediately gives rise to *three* daughter stolons (Fig. 34). One of these is relatively short, and after a fraction of a millimetre suddenly expands to become a bithecal tube. The central daughter stolon is several times longer (sometimes 2–3 mm) but it also eventually expands to become an autothecal (= thecal) tube. The third daughter stolon, some short distance after the establishment of the bitheca and autotheca, again swells to a node and produces three more daughter stolons and the whole process is repeated. What is very neat about this seeming complexity is that it results in each bithecal aperture opening alongside the autothecal aperture *from the preceding nodal division*. Thus if Roman Kozlowski (1948) is correct in supposing that the bithecae and autothecae were, respectively, males and females, then a measure of cross-fertilization was possible by this simple plan of staggering the thecal types. This system is known as triad division of the stolon, and the dendroids typically have the most regular adherence to the scheme.

However, the other groups of Text fig. 1 show much more variety: in graptoloids the bithecae are lost, together with the stolons which connected them all; some tuboids have partly lost their stolons, and have *diad* divisions of the stolons; and others have even greater irregularity *internally*. What is perhaps surprising, however, is that the (auto)thecal spacing remains externally very constant, whatever happens internally to the complexity, or otherwise, of the stipe (e.g. Fig. 45). This is interpreted, in chapter 5, as a direct response to feeding strategies on the part of the colonies: it was necessary to have regularly spaced feeding zooids to efficiently tap the water column; whereas internal complications may be required for other reasons, such as colony strength and even colony shape. Further explanations of the detailed structures of these groups can be found in Oliver Bulman's (1970) Treatise.

Chapter Two

HOW ARE THEY PRESERVED?

Surprisingly little study has been made of the burial processes of graptolites. We generally assume that upon death, benthonic forms, like most of the dendroids, may lie on the sediment surface, their holdfasts (discs or roots, e.g. Fig. 26) still attached to the substrate, as in life. When they were transported by currents before burial, they were first ripped from their holdfasts and the rhabdosomes often become quite fragmented. A complete, or nearly complete specimen, still with its holdfast, is, therefore, a very good indication that the specimen was buried in the same parish where it spent its life. In contrast, the graptoloids, being planktonic forms, sank down through the water column and came to rest on the sea bottom. That is the *shortest* distance they could have travelled prior to burial (see comment on Fig. 36). Bioturbation or turbulent deposition results in twisting or breakage of the rhabdosome. These two modes of life go a long way towards explaining the distribution of the two groups. Dendroids tend to be found relatively sporadically, whereas graptoloids, on the other hand, are far more widespread geographically, consistent with their planktonic mode of life.

In either case, single bedding planes may be found crowded with graptolites, often of only one species or a few at most on a single slab, (Figs 122–124) with a barren interval above and below that may be only a couple of millimetres thick before the next rich horizon, or it may be several centimetres or more. Because of these barren intervals it is important, when collecting, to identify exactly the plane or planes upon which the graptolites rest, and to take care when collecting to split the rock along those planes. In rocks which have not been strongly deformed, this may be a relatively simple procedure, as it is often found that the fossils contribute to a plane of natural weakness and one which may have been further weakened by weathering. In tectonised areas there may be more of a problem, as the rock will tend to split more naturally along the cleavage. In this case it may be necessary to spend some time, first of all correctly identifying the bedding, especially those thin dark laminae within which graptolites tend to be found and secondly acquiring the knack of splitting the rock along the bedding in preference to the cleavage. The graptolites to be found at Abereiddi Bay (mid-Wales coast, Fig. 137 – see Appendix 2) are an instance of fossils needing such a knack which, once acquired, yields a very rewarding collection from an otherwise extremely frustrating locality. This matter is discussed in more detail at the end of chapter 11.

Graptolites are found in a number of sedimentary rock types, but only those of marine origin. Perhaps the most important criterion is the grain size of the sediment, which should be relatively fine, reflecting a sea-bed environment with low, and for the graptolites non destructive energy conditions. Also, fine grains do not disrupt the periderm on diagenetic compaction of the sediments.

The rock type perhaps most typically associated with graptolites is black shale, so much so that a 'graptolitic facies' can be described in which the graptolites are found, often abundantly: muds almost devoid of any other indication of life. Relatively high proportions of carbon, and iron sulphide (pyrite) give these shales their black colour; and they were deposited in anaerobic conditions unfavourable to bottom life (Fig. 24). In such rocks, the interior of the rhabdosome may be filled with pyrite giving a fully three-dimensional internal mould (Fig. 25), which indicates that the pyrite is syngenetic i.e. it developed before the sediment was compacted into a rock. In such cases it is important to realize that the graptolite periderm remains more or less unaltered, with the pyrite moulding the internal surface of the rhabdosomal skeleton. Occasionally a further, patchy layer of pyrite may coat the outside of the periderm, but actual replacement of the graptolite periderm by pyrite is exceedingly rare.

Graptolite fossils are often found in thin layers or even on single bedding surfaces, sometimes in great profusion, separated by unfossiliferous strata generally of somewhat different lithology. Such successions may involve turbidites, which suggests that the graptolites may have been deposited after some bottom current transport down submarine slopes rather than, or as well as, passively sinking to the sediment surface (Fig. 40). It would be important to determine whereabouts in a single turbidite unit the fossils were found, whether within a unit, or towards the top (in the pelagic muds) in order to resolve the question of the possibility of transport.

Soft parts decay, leaving the relatively inert periderm behind, and it is this that comprises the fossil graptolite itself. Fossilised zooids have only once been recorded (Barrie Rickards & Bryan Stait, 1984; Fig. 34 herein) and they were originally pyritised before being weathered to limonite. Electron microscope studies show the periderm to be composed of a number of fabrics each and all made up of fibrils of the protein collagen (Fig. 20) in varied dispositions (Barrie Rickards & Lori Dumican, 1984). However, there are many changes which can occur to the graptolite after burial, the most common of which is the reduction of the protein collagen to a carbon replica.

Overburden pressure causes compaction of the sediment; and it is not unusual to talk of figures of 60 per cent compaction from the original thickness of the sediment, especially in mudrocks (shales) which, as initially laid down, would have contained a high proportion of water. It is quite easy to see that the graptolites must become diagenetically flattened with such a process, but there are problems as to how a highly three-dimensional creature like, for example, the helically spiralled

Monograptus turriculatus could become flattened to a single bedding plane as has occurred at Spengill in the Howgill Fells (contrast Fig. 35 with 37 and 118). It is difficult to see how a thickness of a centimetre or two of sediment (the thickness occupied by the cone of the rhabdosome as it lay on its side during burial) could have become flattened so much as to allow the graptolites to occupy merely a planar surface. Similarly we often find graptolites that have been completely flattened from their original near circular cross-section so again perhaps compaction figures in excess of 60 per cent may have occurred. No satisfactory explanation has yet been given for such preservation (Fig. 118) and in the literature to date the question has never been posed.

In contrast, flattened *Dicellograptus complanatus complexus* from Dob's Linn (Henry Williams, 1981), which is roughly a 'double helix', retains a sliver of sediment between the overlapping portions of the stipes. In other words, it is not flattened on to a single plane, but rather retains some suggestion of the highly three-dimensional shapes of the living rhabdosome. Experimental studies using models could well resolve some of these problems but they are complicated by our lack of knowledge of some critical factors such as the original strength of the periderm.

Carbon has already been mentioned as the preservable elements of which the fossils are now composed. When carbon is present, it is often the periderm itself which has been transformed from the original collagen, by reduction of the original amino acids. Chlorite preservation is a little more complicated. The chlorite (often penninite, an aluminosilicate with iron and magnesium related to mica) forms a coating around the outside of the periderm, covering or disguising many surface details. Graptolites from the Rheidol Gorge, near Aberystwyth, display this feature (Fig. 31); and at Abereiddi Bay the chlorite (weathered whitish or greyish) is a covering over the carbon, which is all that remains of the periderm proper. In some cases the chlorite, or a clay mineral, may replace the carbon. Chlorite is often formed around three dimensional pyritised specimens which have later had the sediment compacted around them: in a 'strain shadow' in fact.

Where graptolites are incorporated in black shales, which are typical deposits of anaerobic environments, the decay of the soft parts promoted the production of pyrite through the activities of sulphur reducing bacteria. This pyrite, a golden yellow or brass coloured mineral (sometimes called Fools Gold, FeS_2), will often crystallise where the decaying tissue was originally concentrated i.e. within the dead graptolite skeleton, producing a typically brassy coloured internal mould. Pyrite preservation may show detail of fusellar rings and the three-dimensional shapes of individual thecae and other internal structures which are not easily deduced by looking at the outside of the rhabdosome (Figs 23 and 25).

On weathering, pyrite oxidises to limonite or haematite and a pyritised specimen will become brown or red and often powdery; consequently the detail of the rhabdosomal structure may be lost. In Devonian monograptids from the Catalonian Coastal Range in Spain, the haematite is enveloped with chlorite which has covered the periderm, leaving little detail other than an outline. Careful removal of this chlorite coating unfortunately also removes whatever remains of the periderm, but it does reveal the haematite internal mould beneath, and on this details of the internal structure become visible, allowing the thecal shapes to be seen clearly and thus facilitating identification of the species.

Nodules form in sedimentary rocks by migration of salt solutions such as those of silica or carbonate towards a nucleus, producing a hard, well-cemented, generally ellipsoidal or sub-spherical mass within the more easily broken sediment. The early cementation of such nodules means that they suffer less from compaction than the host sediment and indeed individual laminae can be traced from the sediment into the nodule, becoming thicker as they enter the nodule. Graptolites which occur in nodules are thus more likely to be preserved unaltered (apart from carbonisation) and to be in three dimensions. It can be very rewarding to dissolve pieces of nodules with acid (see ch. 11 and Fig. 133) and see how well preserved the graptolites are: much detail may still be visible even down to ultrastructural level when seen under the electron microscope.

Graptolites from Cornwallis Island in the Canadian Arctic are found in nodules composed of fine grained calcareous mudstone, and these are amongst the best preserved and most robust graptolites one could hope to find. In contrast, calcareous nodules from Laggan Burn near Moffat in Southern Scotland, while yielding isolatable graptolites (Fig. 131) are somewhat coarser in grain size. The periderm is disrupted and the graptolites in consequence tend to fragment unless great care is taken.

Carbonised graptolites, often with a fill of pyrite or limonite can be found in limestones (Fig. 125). As with calcareous nodules, coarse grained rock tends to disrupt the periderm; this is a problem, particularly where the rock has undergone aggrading neomorphism. (This is where the crystals which make up the rock have recrystallised to a larger grain size). The process of producing larger crystals tends to push periderm aside as the individual crystals enlarge at the expense of other crystals.

Limestones also suffer from pressure solution, where carbonate is removed in solution along a seam, often very contorted and interlocking along the two sides of the seam, producing stylolites. This also tends to destroy potential specimens. Weathering of calcareous rocks may, however, open up planes of weakness parallel to the bedding, and fortunately these are often the very planes upon which the graptolites are to be found. Where these planes weather preferentially, it is much easier to split the rock along the correct planes to find the graptolites. It is, however, frustrating to begin in weathered rock and to find numerous slabs with graptolites, only to hammer back into fresh rock and find that the slabs will no longer split along the right planes! This has happened in the Craven Basin, where there are localities yielding Carboniferous dendroids (some of the last surviving and youngest known graptolites; Fig. 80) – but only in the moderately weathered rock! Too much weathering and a brown, friable 'rottenstone', which completely breaks up on handling, is produced; too fresh a slab and it is impossible to get the rock to split along suitable bedding-parallel planes.

Graptolites of most types occur surprisingly often in various sandstones, both inshore and turbiditic (Fig. 40). Commonly they 'turn up' and are not a viable collecting proposition, having drifted down into the sand bed, after death, perhaps being quite rare. In turbidites they may have been caught up in the flow and are thereafter clastic debris. In such rocks they may occur at the sole of the turbidite, in which case they are strangely undeformed (and often aligned), or they may be within the body of the turbidite, in which case they are more often twisted, bent and broken.

Chapter Three

HOW ARE THEY CONSTRUCTED AND WHAT ARE THEY MADE OF?

(a) PHYSICALLY

Graptolites, when we look at them on a rock surface, or isolated from the rock and floating free in a liquid, impress us either in the bushy form of a dendroid, looking like a small plant in the rock – and they have been and still are mistaken for plants – or by the precise symmetry of a graptoloid – as regular as that of the pearly *Nautilus*. Most dendroids lived attached to the sea floor, and their form, often with a lack of *precise* symmetry, is what we would expect from such a mode of life: like trees in a forest, each specimen has a typical form, but its exact shape is dictated by interference with its neighbours, and probably also by other factors such as the strength and direction of bottom currents (Figs 42–44). By contrast, the graptoloids, living freely in the water column, were not subject to such constraints, and they were able to develop precise forms without hindrance (e.g. Figs 55 and 58).

As colonial animals, each individual or *zooid* of the colony was attached to the others by living tissue. In the graptoloids there is a continuous internal space, the *common canal*, running throughout the colony from the sicula to the growing tips of the stipes (Figs 12 and 23); the dendroids have this space partitioned by diaphragms, but in addition there is a continuous, hollow black cord, the *stolon*, which links all the zooids (Fig. 34). Through this common canal and stolon, and perhaps also through some soft tissue external to the thecae, nervous communication would have been made, and possibly also the transfer of nutrients.

The skeleton of the colony is made up of two main structures, the sicula and the thecae – and nearly all their skeletal elements seem to have been secreted by particular zooids. The initial skeleton was that of the siculozooid, the *sicula*. This started as a larval cup, the *prosicula* – cylindrical in the dendroids, conical in the graptoloids, sometimes boot-shaped or discoidal in encrusting orders. It always appears to be complete when found, and may therefore have been secreted in a single rapid episode of growth. It is sometimes difficult to see on rock surfaces (Figs 10 and 11), but in isolated material it appears to have very thin walls and occasionally just threads. The rest of the colony, including the rest of the sicula, the *metasicula*, grew as a series of increments, all of which have essentially the same form (Figs 13–15).

Dendroid siculae are externally cylindrical because they were attached to some substrate such as the sea floor, shells, or algae, either by a circular or irregular attachment disc (Fig. 26) or by fibrous 'roots'; the larva having settled on the substrate after a period of drifting or swimming. The sicula then grew upwards, before the first thecae grew out of the side of the prosicula by resorbing the periderm to form a hole. By contrast the graptoloid sicula tapers at its closed end into a thin hollow rod or spine, the *nema*. Again the first theca grew out of the side of the sicula, sometimes appearing through the prosicular wall, more commonly through a notch left in the growth increments of the metasicular wall. The aperture of the sicula may be a simple circle, but more commonly it is adorned with either short denticles, or spines (Figs 14 and 50). The most common of these spines, found in all Silurian and many Ordovician graptoloids, is the *virgella* (Fig. 50; and see Chapter 12 for taxonomic significance). It can be longer than the sicula itself, and may have served as the support for the feeding system of the siculozooid (it may be elaborated into a meshwork or rods or lists, and is rarely expanded into a possibly vesicular body). Like the sicula, the thecae may be either simple tubes, with straight apertures, or they may have a complex shape, or have specialised apertures. A recurring theme is the restriction of the aperture: it may face inwards towards the wall of the next theca of the rhabdosome, or simply be constricted so that the shape may be greatly modified, and the area of the aperture much less than the area of the unconstricted theca would have been (Figs 74 and 75). This only happened late in the growth of an individual thecae (growth which is referred to as *ontogeny*). Denticles and spines are also conspicuous features of many thecae (Fig. 76). Paired spines on the apertural rim may have served to support the feeding apparatus, while other single spines may have had a defensive or hydrodynamic role.

All these structures are formed of the two basic types of skeletal material found in the graptolites – the fusellar increments or fuselli, and the cortical bandages (Ch. 1). The fuselli build the skeleton forward; their addition shapes the thecal walls, and they also form the core of the nema and the spines and lists, at least in many graptolites (Fig. 114). The cortical bandages, forming the cortex, thicken up the walls and spines to give strength and resilience to the rhabdosome, and by being added in different thicknesses on different areas provide beautifully engineered structures, as can be seen in the retiolites (Figs 65–71, 97).

When we look at a series of growth stages in a graptoloid, we can see that each theca usually started to grow when the previous one, the 'mother' had almost reached full size, or in some cases rather sooner. Then each 'daughter' theca was in turn the growing end of the stipe, in contrast to some other groups of colonial animals, in which there is a terminal bud, which leads the growth of the branch – this is seen in *Rhabdopleura*, one of the present day hemichordates. When a stipe branches, two thecae are budded off a mother theca. In the early growth stages of many graptolites, there is a complicated series of buds, giving rise to two or more stipes (Fig. 94).

A rather different type of branching is found in the Silurian cyrtograptids (Fig. 58) and similar groups. In these, a theca gave rise to a second daughter theca in the aperture of the parent, which usually had apertural spines one of which then formed the 'nema' of the new branch, and in the same way as the nema itself then grew in advance of the new branch. An analogous structure does occur rarely in some early Ordovician (Arenig) graptolites.

In the early graptoloids, the stipes always grow away from the sicula, while in the later forms the thecae grow along and may enwrap the sicula. In the former the nema is clearly visible projecting from the apex of the sicula (Figs 92 and 96), but in the latter it is enclosed within the two series of thecae, as in the biserial diplograptids (Fig. 107), or runs along the dorsal wall of the single series of thecae, as in the uniserial monograptids (Fig. 110). In both latter *scandent* types the nema grew ahead of the developing thecae.

The retiolites (Figs 65 – 71) are a relatively rare, but absolutely fascinating group of biserial graptoloids, which are found both in Ordovician and Silurian rocks, though curiously the older ones are probably not the direct ancestors of the younger. In a retiolite, part or all of the rhabdosome, as usually preserved, is formed of a fiendishly complicated meshwork of rods or *lists*, with a variable amount of 'ordinary rhabdosome' skeleton as well: the periderm is reduced to a mass of holes! Their appearance, like an extremely fine spider's web, makes them hard to spot on rock surfaces, and so they may be more common than at first appears.

In relatively simple retiolites, some or all of the lists clearly outline the sicula and thecae, but in others the meshwork is more complicated, and includes lists which are not formed in the more usual thecal walls, but represent extra skeletal structures built by the colony outside the thecae, and mantling them. The exact nature of these structures was not established until isolated material was examined in the electron microscope, and though only found in the retiolites, they have important consequences for understanding the graptolites as a whole.

The lists of the thecae are in fact thickenings of the thecal walls, and they extend into normal fusellar tissue. Specimens of some (Fig. 32) show that there is a pale, very thin, wall of material supported by the thickened lists, and often in isolated specimens ragged edges or 'seams' show where the thin walls have been broken away. The lists are formed of concentrations of bandages, or *micro*fusellar parts of normal thecal increments, and owe their strength, and hence preservation, to the concentration of tough cortical fibrils which these give. The fibrils in a list always run along its length, and the result is like a rope, or the cable in a suspension bridge.

The positioning of the lists also adds to the strength, and resilience, of the colony. In *Orthoretiolites*, for example, lists outline the thecal apertures, strengthen their ventral edges, and run across the colony to keep the two sides of the thecae rigid and apart. They are further linked to the nema, which has an identical construction, and as usual, acts as an axis to the colony. As a result of this the colony is, even as a fossil, more resilient than an 'ordinary' graptolite – when well preserved they are springy when handled. This lightness with resilience may have enabled them to colonise more turbulent waters than other graptolites.

In the Ordovician retiolites, *Phormograptus* and *Pipiograptus*, the framework of thecal lists is enclosed in an outer meshwork increasing the volume enclosed to three or four times that of the thecae themselves. These lists are unseamed, and so cannot have been formed in any skeletal walls, and they form a three dimensional meshwork. In specimens of *Phormograptus* the lists are stouter than those of the thecal walls inside, and so they appear to have taken the main burden of supporting the colony.

In the Silurian retiolites, such as the genus *Retiolites* itself (Fig. 70) the outer meshwork *is* seamed (as in Fig. 68), and the lists make a sheet linked to the thecal apertures. Before study in the electron microscope, it was thought that this mesh was in fact part of the thecal walls, but it has now been established that it is external to them. The seams indicate that in this case the mesh was continuous in life, with a very thin periderm of now-degraded fuselli forming the surface, on which the lists were laid down, again as bandages. When these graptolites are reconstructed, it can be seen that the outer sleeve enclosed two canals running the length of the colony, linked to the thecal apertures, and to other apertures at the proximal end belonging to the sicula, the distal end, and in some cases a series of aperture-like openings along the midlines of the colony (as in *Stomatograptus*; Fig. 71).

In *Retiolites* the lists of the thecae are more robust than those of the outer sleeve, but in later retiolites the reverse is true: in forms like *Gothograptus* (Fig. 67) the thecal framework is very frail, and difficult to reconstruct, the preserved material being almost all part of the sleeve. Here, as in *Phormograptus*, the outer extrathecal framework has taken on the support of the colony, and the original thecal walls need not have any strength.

What was the purpose of these peculiar structures? They were evidently not initially developed to support the colony – the stratigraphical record shows that that was a later role. The sheet-like nature of the mesh in *Retiolites*, and the system of canals produced, indicate that it was not a 'buffer zone' protecting the thecae inside from damage: the canals were important. In the absence of any firm evidence of the nature of the graptolite zooids, we can only speculate that the canal system had perhaps a feeding function. A system of afferent or inhalent currents, and efferent or exhalent currents, may have been set up, probably by ciliary action within the canals, and this could also have contributed to automobility of the graptolite. In forms like *Stomatograptus* the directed currents may have been inwards to the thecal apertures, and outwards from the side openings or stomata, and the net effect may have been to have kept the graptolite 'hovering' as it received both nutrients and oxygen. One of the major debates in graptolite research is about whether graptolites *were* automobile or whether they were passive responders to the vagaries of the ocean's currents.

Phormograptus may well have had a similar system of canals between the lists of its jacket, but unfortunately without seams we cannot reconstruct any enclosures.

However, there are larger openings in the meshwork opposite the thecal apertures, which does support a similar reconstruction. One can imagine a system, formed within soft tissue supported on the meshwork.

(b) BIOCHEMICALLY AND BIOLOGICALLY

It is clear that the graptolite colonies were constructed physically in a varied and often complex fashion as they evolved into niches we cannot as yet identify. Our understanding of exactly how the zooid secreted the skeletal layers is not fully agreed: it is the subject of yet another debate. However, what we do now know is that the whole of the periderm, including all spines and processes, is made of the protein collagen. Collagen is a fibrous protein which takes slightly different physical form to suit different functions. We have already mentioned that fusellar fibrils have a spongy texture; cortical fibrils have a texture of parallel bundles or rope-like arrangement (Fig. 20); and some especial strengthening rods have a triple helix coil even within a single fibril. We know that all these structures are biochemically the same because they grade into one another, and are admixed one within the other, rather frequently. Therefore the secretory mechanism which deposited one form of collagen is likely to have secreted the remainder. In all probability the very same secretory cells were also used in each case.

Because modern rhabdopleurans can do exactly the same thing it has been argued on this and on other grounds that a rhabdopleuran secretory model holds for graptolites too (Peter Crowther and Barrie Rickards, 1984; Barrie Rickards and Lori Dumican, 1984). The debate arises because this thesis is challenged (e.g. Nancy Kirk and Dennis Bates, 1986) and a mechanism of secretion by zooids connected to extrathecal tissue has been proposed. The drawbacks to both ideas are outlined in Ch. 1. What is interesting is that despite the enormous advances in understanding the biochemical origin of the skeleton, we still have not solved the biological secretory mechanism of the zooids.

Chapter Four

WHERE ARE THEY FOUND?

(a) GEOGRAPHICAL DISTRIBUTION

Graptolites have been found on every continent except Antarctica. Their time distribution group by group is given in Chapter 1 and it follows that they will occur and potentially can be found wherever Cambrian to Carboniferous rocks crop out. However, they are relatively uncommon at both ends of their range in the Cambrian and Carboniferous, and most common, often very abundant, in Ordovician and Silurian rocks. In certain regions they occur frequently in Devonian rocks, too.

Cambrian graptolites – entirely bottom-living – have been found most frequently in Australia, North America and the Soviet Union, with some groups also occurring in China. They have a tendency to occur in relatively shallow, shelf sea deposits, in calcareous beds and sandstones, often in association with trilobites, crinoids and other benthic fossils. They are rarer in offshore, deeper, black shale deposits, and when they do occur in these deposits may well have been carried there by currents from an inshore, benthic source.

If we may discuss now their latest time occurrence, namely those graptolites in Carboniferous strata, it would seem that as in the Cambrian all the records are of benthonic types. Graptolites are rarest in the Carboniferous, but they do occur in North America, China, and in Europe (on the mainland in Belgium, but also especially in Northern England; Fig. 80). It is still likely that many Carboniferous graptolites have been overlooked because of their superficial resemblance to fenestellid bryozoans. Because of this close similarity we are frequently sent graptolites which turn out to be bryozoans.

Graptolites do not, on present evidence, survive the Carboniferous: one Chinese claim from the Permian we do not accept because the evidence of age is not clear, though the specimen is undoubtedly a graptolite. Because graptolites were approaching extinction in the Carboniferous it is likely that they were somewhat restricted geographically and possibly declined in numbers.

Ordovician and Silurian times saw the acme of graptolites, especially planktonic forms, and these were extremely widespread occurring almost everywhere where rocks of these ages crop out. Whilst they are found in inshore deposits or shelf sea deposits, along with benthonic graptolites and other benthos, they also sank upon death into offshore mudrock sequences giving rise to a typical black, graptolitic shale facies. Famous occurrences of such facies are found in many parts of the world: from the South American Andes to the Mackenzie River at the northern end of the Rocky Mountains and Cornwallis Island in the Canadian Arctic Archipelago; in Britain from Wales to the Lake District and the Southern Uplands of Scotland; in Northern Europe on the Island of Gotland and in Skane in southern Sweden; the Bohemian basin of Czechoslovakia; the Ural Mountains and parts of Siberia; many parts of China and elsewhere in Asia; from New South Wales and Victoria in Australia and in New Zealand; the Atlas Mountains of Morocco in North Africa (but not yet in Southern Africa) and from Sardinia and Turkey in the Mediterranian region.

They seem to have been more restricted geographically in the Devonian, tending to persist where the off-shore marine black shale persisted as a facies: thus they have not yet been found in marine Devonian strata (and late Silurian strata) in the UK but do occur in Central Europe. Similarily, distributions in the Americas, the Soviet Union, Asia, China and Australia are progressively restricted through time. Nevertheless a great deal of work has been carried out on Devonian planktonic graptolites, though it is one of the main areas of research where there is yet a great deal to do.

(b) STRATIGRAPHICAL DISTRIBUTION

The earliest graptolites are found in the Middle Cambrian. These were the dendroid graptolites, filter-feeders which lived attached to the sea floor. This niche was even then a fairly crowded one. Living space and resources were shared – and competed for – by sponges, echinoderms, pennatulids and brachiopods among others. The graptolites seemed destined to be a small and not terribly significant part of this fauna. But a single evolutionary step in the earliest Ordovician Tremadoc Series liberated one population to graze among the phytoplankton of the high seas. Probably the planktonic larvae of that population simply did not settle down to a benthonic lifestyle. The planktonic graptolites had arrived.

The first few million years of their history were marked by rather drastic changes in design. This is not at all surprising. The first planktonic dendroids were creatures designed for life on the sea floor, not for the open ocean. A huge new ecological niche had opened up to them and they coped, indeed prospered; but there was room for a great deal of structural improvement, after a presumably successful genetic preadaption.

Thus, in the Tremadoc, design faults were quickly ironed out, as these graptolites competed among themselves to find the mode of colony construction best suited for a planktonic life. Most significantly, there were dramatic increases in colony size: the number of branches fell from hundreds to sixteen, to eight and, by the beginning of the Arenig, had stabilised at, usually, two or four (Figs 81–96).

A major internal 'domestic' rearrangement also took place within the colony. The rooted dendroids and early planktonic dendroids possessed two kinds of thecae, thought to house male and female zooids respectively (Ch. 1). At the beginning of the Arenig, one type (termed bithecae) had been largely discarded; all future planktonic graptolites were to be, like the common garden snail, hermaphroditic. These are the graptoloids.

Evolution among these early Ordovician two- and four-branched graptoloids – the dichograptids – generally took the form of varying the relative attitudes of the branches (Figs 88–96). Among the early two-branched dichograptids, the tendency was to go from horizontally outstretched branches to, later in the Arenig and in the Llanvirn, parallel branches hanging down from the initial bud (the 'tuning-fork' graptolites). In the deeper waters of the palaeo-tropical latitudes (now Australia, and North America) thickened branches in a V-shaped attitude (the isograptids) were common (Fig. 18). The four-branched dichograptids did not show such clear trends; but one among the possible attitudes was to direct the four branches upwards, back to back. This was a foretaste of the graptolite of the future – though not, it is thought, a direct ancestor: the leaf-like *Phyllograptus* shape so formed was certainly advantageous enough to have been arrived at quite independently along two separate lineages. Roger Cooper and Richard Fortey (1982) dissected at the early-formed (skeletal) innards of sundry 'phyllograptids' of Arenig age, and were able to discern two radically different mechanisms of early development. Thus the pseudophyllograptids were recognised, and took their place as a rather fine example of parallel evolution alongside the phyllograptids proper.

The thecal tubes of dichograptids tended to evolve rather unadventurously. For the most part, they stayed as simple tubes, becoming longer and thinner in some species, and shorter and fatter in others. Exceptions to this conservatism have been found in the early Ordovician strata of China and some other regions – the appropriately named sinograptids possess thecal tubes corrugated into folds of impressive amplitude (Figs 10 and 11). Another single exception constitutes one of those seeming examples of punctuated equilibria in evolution thrown up by the fossil record. The tuning-fork shaped *Aulograptus cucullus* has the distinctive sharp, right angle bend of the climacograptid thecal type; no fossils have been found to bridge the large morphological gap between this common species and other dichograptids.

Take four graptolite branches and fuse them together back to back and you get a recipe for shortlived success. Do the same to two branches and this success becomes considerably greater and longer-lasting. Thus the diplograptids which appeared near the end of the Arenig, probably via more than one evolutionary lineage, formed a large part of the graptoloid fauna in the Llanvirn, and from the mid Ordovician to the early Silurian, more or less took over the high seas. Whatever problems of 'feeding space' might be assumed to have taken place by packing the zooids more closely together, they were decidedly outweighed by the advantages given by a symmetrical, compact colony.

These advantages remain obscure to us, but are fun to discuss (see Ch. 5). The diplograptids enjoyed more success over a greater length of time (ca. 60 million years) than did any other group of graptoloids. Moreover, they did not appear to have needed any great morphological changes over this time to maintain their dominance, as other forms of graptoloid came and went. Within the great majority of diplograptids there was a relatively limited range in size, a few popular but unspectacular modes of constructing thecal tubes, and a dozen or so methods of initial 'embryonic' development (now thought to be one of the more reliable methods of working out which graptoloid may be related to which; Ch. 12).

There were relatively few exceptions to this diplograptid 'norm'. The genus *Cryptograptus* constructed its skeleton in a rather complicated fashion, by wrapping the thecal tubes sideways around the main colony axis; it seems to have been an early parallel attempt to form a diplograptid skeleton. The retiolitids possessed a skeletal meshwork, rather than a solid-walled skeleton. These were the longest-lived of the diplograptids, being represented well into the Silurian (Ludlow Series). Rather mysteriously, there are rather large gaps in the retiolitid record, and one of the more attractive explanations of this has lineages becoming totally soft-bodied at certain times.

During the Llanvirn, the diplograptids shared dominance with the pendent, 'tuning-fork' dichograptids. More or less coincident with the demise of these latter midway through the Ordovician, a wave of new graptolites arrived more or less simultaneously, and more or less worldwide. This was the '*gracilis*' fauna: new colony shapes and new forms of thecae appeared: *Nemagraptus gracilis* itself, elegantly S-shaped, with two long fringes of secondary branches; the V-shaped dicellograptids (Fig. 27); and the Y-shaped dicranograptids. These latter constituted one of the best examples of precision design in graptoloid skeleton construction – with tightly-controlled complexity in both overall colony form and thecal shape (arguments about the purpose of such precision are aired in Ch. 5). There were also the thorn shaped corynoidids, which took the trend of colony reduction almost to vanishing point, with nothing left but a sicula and a couple of downward growing thecae, plus a seemingly aborted third theca.

The *gracilis* fauna enjoyed a lot of success initially, and some elements (the dicellograptids) persisted until late in the Ordovician. The diplograptids, though, tightened their grip on the planktonic realm throughout the upper part of the Ordovician. Again, they seemed to have achieved this without resorting to anything strenuous in the way of morphological evolution; some forms grew spines, and this may have been an adaption to retard sinking of a now very compact colony.

The diplograptids were equal (but only just) to the global environmental crisis of the end-Ordovician. The nature of this crisis is now quite familiar to us. Africa, in the course of its slow migration across the earth's surface, had crept to a position over the South Pole. Ice-sheets formed, and a full-scale glaciation began. If analogy with the current glaciation can be made, ocean temperatures in high and mid latitudes would have dropped sharply, restricting the area of warm tropical water to a narrow belt around the equator. As sea level fell, the area of shelf sea was drastically reduced.

The class Graptolithina survived, but it was a very near thing. Literally just two or three graptoloid species – all diplograptids – hung on until the north African ice-sheets began to recede at the beginning of the Silurian. From those few remaining species was to come the next major breakthrough in colony design, and an evolutionary explosion.

As evolutionary explosions go, though, it was a little slow in igniting. The first representative of the new order appeared in the end-Ordovician, but as such an elusive and obscure element of an overwhelmingly diplograptid fauna (Fig. 98) that it was not discovered until a couple of decades ago, even within the classic, intensively collected Dob's Linn section in Scotland (see Appendix 2).

The evolutionary step was, to outward appearance, a simple one. One of the rows of thecae disappeared, leaving a single row of thecae (Fig. 103) growing upwards from the sicula. Thus the monograptids arose. In terms of developmental mechanics, though, the step was nowhere near as straightforward as merely suppressing the growth of a single thecal row. The untangling of the rather complicated twists and turns of the early thecal development of the diplograptid colony was also required. The step seems to have been difficult not only in terms of altering the colony's development: the scarcity of the earliest monograptids indicates that there were teething problems with the maintenance of uniserial planktonic colonies, and these took a little time to overcome.

Indeed, for the next couple of million years or so, monograptids continued to be dominated by the diplograptids. The dimorphograptids appeared, though (Fig. 108), and became common. These, graptoloids which start off uniserial before adding a second row of thecae later in development, once appeared to represent a highly satisfactory 'missing link' between the diplograptids and the monograptids. Alas, nothing of the sort. Dimorphograptids arose from early diplograptid ancestors; thus, they rather represent failed monograptids which 'chickened out' in mid-development, reverting to the familiar, comfortable diplograptid mode!

Soon afterwards the monograptids grew rapidly in abundance and variety, and by the middle to early Llandovery times, they had firmly supplanted the diplograptids as the dominant group. Whatever problems the adoption of a uniserial mode may initially have caused, it allowed a far greater range of both thecal and colony shape than the diplograptids were capable of. Early expressions of this plasticity were the evolution of lineages with strongly curved colonies (Figs 36 and 79), and the elongation of the thecae at first into a high, triangular shape (Fig. 74), this elongation then being taken to extremes among the curious thread-like rastritids (Fig. 55), which became abundant in the mid-Llandovery. The diplograptids also tried a new tack at about this time, developing flat, tabular colonies – the petalograptids (Fig. 23) – within which one particularly adventurous lineage developed the longest thecae known, each thecal tube being up to 30 mm long. Some rare forms of this group exhibit what is almost a non-colonial development, comprising a sicula and one theca only.

In the later Llandovery times the monograptids in one sense suffered a decline – there was a drop in species diversity – yet innovations in colony design continued apace. Colonies curved into conical spirals were now common, while thecal tubes assumed quite fantastical shapes, becoming twisted or even coiled, either in the plane of the colony or sideways, with spines here and there for added baroque effect. The reason for this burst of contortionism is not clear; being unscientific about it, it looks as though the zooids wanted to hide. Not all, though, the contemporaneous pristiograptids appeared to be of less paranoid persuasion, possessing straight thecal tubes with open apertures of absolute simplicity. The most abundant monograptids of this time took a middle course between these extremes, with thecal tubes bent over into hooks.

By the very end of the Llandovery, the continued drop in species diversity had become something of a crisis. The decline was almost as serious as during the end-Ordovician, except this time there was no obvious external factor that we yet know of, no environmental 'smoking gun' to provoke it.

Things improved in the Wenlock, with the proliferation of the exceedingly graceful cyrtograptids (Fig. 58). This family of graptoloids had made a quiet entrance in the late Llandovery, with one lineage of spirally coiled graptoloids reversing the general trend towards smaller colonies by adding secondary branches (cladia). The rest of the Wenlock graptoloids were, one has to say, rather a sober and conservative lot. Monograptids with an L-shaped kink (or geniculum) in the thecae were prominent early (Fig. 29), those with hooked and simple tubes also being common. An intriguing subtlety is associated with the latter. Unlike the otherwise similar pristiograptids of the Llandovery, major lineages in the Wenlock had a slight – but very persistent – curve in the early part of the colony. An adaptive feature of hydrodynamic importance, (see Ch. 5) or a feature of such profound irrelevance to the colonies' survival that they didn't bother to evolve it away? The question is open – as are many of this type in palaeontology.

Another apparently unprovoked crash in species diversity came near the end of the Wenlock era. For a time it seems that there were only two species of graptoloid living worldwide (though one or two more must have been lurking in unrecognised spots). The retiolitids just survived this late-Wenlock crisis, which may have been related to the global marine regression of the time. Recovery in Ludlow times, though, was marked by the continued dominance of monograptids. These mostly had simple thecae – though yet another slight, but very persistent, morphological trait appeared – the presence of pairs of small spines, generally confined to the first two or three thecae (Fig. 110). A few lineages took up the trend of thecal elaboration, which had been largely absent during the Wenlock. In some instances this was carried to extremes; the bizarre asymmetrical thecal tube of a later cucullograptid looked like an early evolutionary forerunner of the less presentable ear of an unsuccessful heavyweight boxer.

The last surviving graptoloids are found in the early to mid-Devonian. Evolutionary change was evident right to the end; it was not a case of the last graptoloids being simple forms which hung on before fizzling out. Two fairly complex forms of 'cyrtograptid' were present, together with monograptids, now mostly characterised by hooded thecae giving way later in colony growth to sharply kinked (geniculate) thecae capped by hood-like structures.

This relatively diverse graptoloid assemblage disappeared fairly suddenly. The reasons for extinction remain obscure. Crudely speaking, the downfall of the graptoloids coincided with the spread of fish into the planktonic realm. This correlation, though, is terribly simplistic, if not downright misleading. Perhaps the final extinction was simply akin to the early Silurian species diversity crashes, but carried through to finality. Also these early jawed fish were in no shape to function as graptolite predators, mostly being extremely unwieldy, shoreline-hugging creatures which would not have had the faintest notion of what to do with a graptolite even in the unlikely event of them catching one. Other suggestions have been made for the extinction of graptolites, Tania Koren' and Barrie Rickards (1979), for example, suggesting a link with the changing land plant developments, and global palaeogeographical changes.

The dendroids, though, did not yet become extinct. They had remained more or less unchanged, as a minor element of the fauna on the sea floor throughout the Silurian and Devonian, while the graptoloids went through their dazzling evolutionary changes. The dendroids persisted for a further 50 million years or so, finally dying out in the upper Carboniferous. The related pterobranchs are an even finer example of the ultimate longevity which a shy and retiring lifestyle can give. These dendroid-like hemichordates originated in the Middle Cambrian too, together with the graptolites (Text fig. 1). They then dropped out of sight in the geological record even more successfully than the coelocanth, being recorded only at wide intervals of time. Reported sporadically throughout the Phanerozoic from the Ordovician, Silurian, Cretaceous and Eocene, today they can be found hiding under upturned shells in a few of the more obscure Norwegian fjords, upon sea mounts and in some estuaries, and close to tropical shores. The species seem unusually tolerant of depth and temperature changes. Although these living pterobranchs were first discovered over a hundred years ago now many aspects of their biology are still not understood. This applies especially to their life habits, such as the construction of their skeletons. The apparent neglect of these curious little animals by zoologists may be partly due to the difficulty of observing them *in vitro* since they react negatively to any disturbance be it heat, light, vibration etc. consequently they can only be investigated under very carefully controlled laboratory conditions. Fortunately in recent years Professor Noel Dilly, an eminent medical researcher, has diverted some of his many talents and energy to tackling these problems and enlarging our understanding of these extraordinary survivors of an evolutionary radiation that took place over 550 million years ago in the Middle or possibly Lower Cambrian. It is a good thing for palaeontology that they did survive, albeit in such a diminished way; without them, interpreting the graptolite skeleton (Chs 1 and 3) would have been an even more perplexing occupation.

CHAPTER FIVE

HOW DID THEY LIVE?

(a) GENERAL LIFE MODE AND PASSIVE RESPONSE VERSUS AUTOMOBILITY

Graptolites are found almost all over the world, in limestones, sandstones and shales which were deposited in a wide range of environments. They are sometimes found associated with trilobites, brachiopods, hydroids, chitinozoans, sponges, foraminifera, dinoflagellates and acritarchs. Yet the fundamental questions of how they lived remain largely unanswered.

The problem is that not only are they extinct, but also that they have few living relatives. The ones they do have, the hemichordates *Rhabdopleura* and *Cephalodiscus*, live in a limited range of benthic environments, which can only help to explain by analogy the life habits of a few of the bottom living graptolites. More difficulties arise from the general absence of preserved soft tissue. Soft tissue might have completely covered the rhabdosome. Alternatively, it might have been confined to the zooids in the thecae themselves, or have been present on some of the vanes and webs which are sometimes developed. Graptolite zooids have only once been found preserved and there is no indication there of extrathecal tissue.

However, much is known and remains to be discovered from the hard part structure which often survived diagenesis remarkably well. Each structure must have evolved for a particular purpose, and many of these can now be suggested with a fair degree of confidence.

Of the eight main orders of graptolite, most of the members of seven were benthic in life habit. That is, they lived attached to the seabed like their living relatives. These are the orders Crustoidea, Camaroidea, Stolonoidea, Tuboidea, Dendroidea, Dithecoidea and Archaeodendrida (Fig. 1). Several of these are extremely rare. The camaroids, and crustoids were encrusters, living firmly attached to a surface and moulded to it. The attachment surface is flattish, with a membranous layer for sticking down the earliest growth stages of the colony. This probably means (according to Barrie Rickards, 1976) that the larva attached itself before developing a skeleton, which grew with reference to the nature of the substrate.

The dendroids are much more common as fossils and although most seem to have been benthonic, they were not encrusting but supported themselves above the seafloor by holdfasts beneath a more robustly-secreted skeleton. They are usually found in association with other benthic species of trilobites or brachiopods, and in relatively shallow, inshore facies rocks – coarse sands, silty mudstones (even conglomerates) and limestones of various kinds. They have basal discs or root-like holdfasts (Fig. 26) which may have secured them against currents. The discs vary in form, seemingly due to the variable nature of the substrate. There were adaptations to hard bottoms and rooting fibres which provided a better grip on soft sand and mud.

Although most dendroids lived like this, some evolved to inhabit the planktonic environment, and finally evolved into the graptoloids. These forms, like *Rhabdinopora flabelliformis*, had no holdfast (Fig. 126; Fig. 41 is a similar species). Instead they sometimes had a float-like structure at the proximal end, or simply a straight or divided nema. They are found in many different kinds of sedimentary rocks including black shales, which were deposited in anaerobic conditions on the seabed. In this environment little benthos could have lived as there would have been little or no oxygen available for respiration. The planktonic forms are found over a greater geographical area than their attached relatives and were, in fact, more or less cosmopolitan. Oceans had ceased to be a boundary to their migration – they were part of the plankton rather than of the benthos.

Planktonic animals float freely in water without the ability to swim against currents. Modern members of the zooplankton fall into three main groups. The first are larval stages of benthic adults; the second, known as epiplankton, live attached to floating seaweed or on the organic aggregates known as sea snow or on buoyant debris; the third are members of the true plankton which support themselves in the water by adaptations to their own body plan. Graptoloids and the planktonic dendroids are obviously not larval phases and for a long time they were believed to have been members of the epiplankton. In 1897, Lapworth suggested that they had attached themselves to floating seaweed, via the nema, and lived underneath 'like a bell at the end of a rope'.

The view that all graptoloids and planktonic dendroids were epiplanktic has been shown to be wrong as more has become known about graptolite morphology. Bulman (1964) showed that in some graptolites, such as some *Dicranograptus* and *Phyllograptus*, no nema has ever been recorded. In others the nema is much shorter and thinner than the size of the rhabdosome would seem to require for secure attachment. In others still, like *Monograptus turriculatus*, the nema seems to project at entirely the wrong angle relative to the rest of the rhabdosome. More importantly the nemal tips show no indications whatever of modifications for attachment (contrast holdfasts of the dendroids, in Barrie Rickards' 1976 paper) so that any contact for an epiplanktonic mode could only have been a soft tissue connection. Bulman concluded that the planktonic graptoloids were members of the true plankton. This is now the accepted mode of life for the planktonic dendroids and the Order Graptoloidea which evolved from them, all of whose members are planktonic.

The implication of such a conclusion or hypothesis regarding their mode of life is that structures evolved in the graptolites must have been designed to aid or at least not hinder planktonic life, with its problems of retaining or controlling vertical position in the water and of finding enough food while moving more or less helplessly through a liquid medium. Attempts to understand graptolite modifications in this light have been going on for many years and much progress has now been made although disagreements still remain.

The opposing views that have been postulated for graptolite mode of life are basically incompatible, and deal partly with assumptions which are untestable at the present. It is logical to treat each view separately and then to consider some recent studies which may finally suggest a kind of synthesis of views on graptolite mode of life.

On one side of the argument is the view put forward over many years by Oliver Bulman, Barrie Rickards and others. They advocate a passive mode of life with graptolites floating at various levels in the water column and being moved by currents and eddies. They would have developed from a planktonic larval stage into a planktonic adulthood. They would usually have been orientated with the sicula pendant from a nema, but they might well have controlled their position in the water column by controlling fat body secretion or gas body secretion, thus moving slowly up or down.

Vanes are common, particularly in biserial graptolites and can be ribbon-like, two or three bladed (Figs 24 and 107) or a mass of fibres (Fig. 106), and there are a few that may be vesicular bodies. They would have projected above the colony and have been used to aid buoyancy. They were possibly coated with a layer of extrathecal tissue which might have been gas-rich (i.e. vacuolated tissue). Alternatively, increased surface area alone might have been sufficient to retard sinking (Charles Mitchell and K. Carle, 1986). The colony might have been able to change its density in order to move up or down through the water column. In spiral shaped colonies, for instance the cyrtograptids and *Monograptus turriculatus*, movement would have been spiral as well, induced passively by the shape of the rhabdosome responding to turbulence.

Many graptolites might have lived at a high level in the water, and scandency (having two rows of thecae joined back to back) might have evolved in order to protect the delicate proximal part of the rhabdosome from turbulence right at the surface. Such scandency produced a relatively thick and dense structure on which thecal spines and processes commonly developed to retard sinking. Webs which are common in planar, multiramous graptolites would also have retarded sinking. They might also have served as attachment sites for buoyant extrathecal tissue.

An alternative view of graptolite life habit has been mainly put forward by Nancy Kirk, over the last twenty-five years. She envisages graptolites with a benthic larval stage even when the adults were truly planktonic. As a consequence of this early attachment they would have floated with the sicular aperture up, but would have needed to control this orientation by concerted zooidal activity. Thus vanes would have been situated below the colony and provided a weight to maintain orientation. They would also have stabilized the colony, preventing it from swaying too and fro in the currents. It should be said that there is strong evidence that the larval stages did *not* attach, both from the morphological and from the sedimentological standpoint.

The individual zooids would, on Nancy Kirk's model, have been capable of producing feeding currents of a sufficient strength to propel the colony. This obviates the need for many buoyancy structures. Because of their early fixed life habit they would have tended to live at depth in the water column. Many adaptations of rhabdosome morphology relate to feeding efficiency; many planar, multiramous shapes were efficient at tapping a circular area of water. Large or delicate rhabdosomes, or those with spines and other processes would have had any movement through the water impeded. Cyrtograptids and monograptids with any degree of curvature, not just spiral forms, would have had spiral feeding paths, caused by zooidal propulsion working on the rhabdosome.

Thus one of the most basic disagreements between graptolite workers is whether the graptolites were automobile or retained their position in the water column by passive means. A passive mode of life is suggested by the small size and radial arrangement of the zooids, in many forms, which are the living parts of the colony and hence the only source of generated currents. The only likely source of current generation in a zooid would be its feeding structures, and these would have been specifically designed to allow the maximum possible amount of water to pass through. In modern members of the plankton, movement by muscle power is the only efficient means of propelling relatively large organisms, comparable to the size of a graptolite. An *automobile* life is suggested by the fact that currents of a kind are generated by modern filter feeding benthos, which might help to propel them if they ceased to be attached to the substrate. Most living members of the zooplankton are automobile with varying degrees of efficiency, although some regulate their position in the water by passive changes in the buoyancy of the colony, and diurnal migration, up and down, has developed independently in several planktonic groups.

The two opposing hypotheses remain untestable, but to some extent need not colour our interpretation of many structures. Both models assume that the colony moved in some way relative to the surrounding water so that the net result was the passage of water over the rhabdosome. This movement could have been achieved either by automobility or by passive density changes. The only unacceptable hypothesis is that of passive, neutrally buoyant rhabdosomes which would have floated with the same body of water until they starved to death! But no one is suggesting this scenario.

This unexpected agreement allows real models of graptolites to be built and then tested to see how they react in water. The models are built to sink passively, but this could equally be regarded as an approximation of active swimming.

Bulman, Bates, Kirk and Rickards all agree that some rhabdosomes are spiral in form and probably rotated while moving through the water. These include the cyrtograptids, *Monograptus turriculatus* and some spiral dicello- and dicranograptids.

Recent work with graptolite models suggests that *many more* shapes were also designed to rotate. Vanes on biserial forms, and dorso-ventral curvature of some monograptids is now known to cause rotation. So does the offset of thecae around the stipe of dicellograptids (and presumably of dicranograptids as well). This modifies the stipe into a twisted ribbon shape which causes rotation of the models as they fall through the water. This is regardless of the shape of the colony as a whole, so that even non-spiral dicellograptids rotate. Planar, multiramous forms (Fig. 2) can be made to rotate very easily, by slightly offsetting each stipe relative to the next to give a propeller shape, an arrangement suggested by the actual preservation of some dichograptids.

Such rotation was probably necessary for two reasons. First, all colonies would have needed to maximise the amount of water through which each zooid fell. However, movement through the water would have required an expenditure of energy, either by zooid activity or by buoyancy changes, to return the colony to its original position. Passive rotation would have increased the feeding path of a zooid without increasing the distance of movement. Secondly, vertically disposed graptolites – the monograptids and fused biserial forms – have zooids vertically stacked one above the other. They would, therefore, have had the problem of preventing water depletion by one zooid passing over the next. This would not happen if each zooid had its own rotating path.

There are other ways in which graptolites seem to have adapted to a maximum efficiency food gathering strategy. The branching pattern of multiramous forms seems to have depended on achieving the maximum possible feeding area without any competition between zooids. Richard Fortey and Adrian Bell (1987) have produced computer simulations of graptolite colony plans and have found that they can be generated very simply, at least in two dimensions, with a small number of instructions about frequency of branching. The patterns seem to approximate to theoretically determined optimal harvesting arrays (see next section, this chapter). Even more interesting is the fact that in the computer models, branches eventually tend to overlap if the branching instructions are repeated enough times. In real colonies however this does not happen; if necessary an expected branch is suppressed. Not only does this show that competition within the colony was to be avoided, it also confirms the overall genetic control of colonial development for which there are now several independent lines of evidence. The state of the art in understanding how graptolites lived seems to be as follows; two classical ideas of graptolite mode of life have developed over the last thirty years or so. They are basically opposed and contain many untestable, though not unarguable, assertions. However, both have been accepted by many people as suggesting a reasonable model of what graptolites did. Recent work has tended to concentrate on testable aspects of these two models. Computer and real models of graptolites offer new ways of acquiring information when the obvious sources available for other groups of fossils are missing.

Almost incidentally, some of the main reasons for graptolites to have developed in the way they did, are emerging from this work. Maximum feeding efficiency seems to have been a driving force. This was achieved by efficiently covering the circular area of the colony with zooids, maximising the area of water through which they fell. Another control seems to have been effected by the need to reduce sinking rate to a minimum through the development of webs, spines, vanes and lacinia. This was presumably because of the relatively 'high cost' of the energy needed to return a rhabdosome to its original level in the water. This links neatly in to a tendency to rotate during movement which allowed a maximum amount of water to be sampled with minimum vertical displacement of the colony.

All graptoloids and some dendroids were planktonic. All other graptolites were members of the benthos. The benthic forms lived in shallow water and filter fed from the water that moved over them. The planktonic forms adapted to maximise their feeding area and to achieve a reduced or variable buoyancy. This was advantageous either in order to move passively or via self propulsion through the water. What they ate is unknown although it is assumed that they fed on the surrounding phytoplankton, and perhaps on small zooplankton and detritus. More details of life habit are constantly emerging, but at the moment the picture remains vague and tantalising. At present it seems that no theory should be rejected out of hand, and that an ultimate synthesis, if it ever becomes possible, is likely to incorporate elements from all of the sources considered here.

(b) RHABDOSOME PATTERNS FOR HARVESTING: GRAPTOLITE COLONY SHAPE

One of the most appealing features of many-stiped graptolite colonies is their remarkable regularity. They produce the most precisely organised colonies of any of those groups of organisms with colonial habits, including corals, hydrozoans or bryozoans. These regular colonies may range in size from a centimetre to over a metre in diameter. It is, unfortunately, very difficult to collect such giant colonies because graptolitic rocks are usually too friable or cleaved. Even if you can see a giant colony in the field the rock may fall to pieces if you try to collect the whole. The numbers of individual zooids in some of the larger colonies certainly numbered several hundreds in the graptoloids and several thousand in some dendroids.

Quite often the early parts of the larger graptoloid colonies became secondarily thickened with cortical tissue (Fig. 83). In these, it is likely that the zooids in the first few stipes were atrophied or dead, perhaps overgrown by cortical tissue, and that the living, feeding zooids were concentrated on the outer parts of the colony. Because we know that the graptoloid colonies of this kind were planktonic, it seems entirely sensible to explain the colony form in terms of adaptation to the open ocean habitat. The regularity of the colonies compared with encrusting or reef-forming colonial organisms may in part be a product of freeing the graptolite from the constraints of the substrate, where any kind of obstacle, or the vagaries of ocean currents, or the influence of neighbouring organisms can influence the final form of the colony. Graptoloids were free to grow outwards from the first individual of the colony unencumbered by such constraints. The size of the colony was partly a species characteristic, but there is some evidence that certain species could go on growing, getting larger and larger and adding more branches as they grew, until they reached giant proportions. Quite how fast they grew is a matter for speculation, but it does seem likely that the giant colonies must have been several years old, for the secretion of an individual growth ring in modern hemichordates takes eight hours: there are many thousands of such rings along the stipe of a giant graptoloid.

Branched graptolite colonies are so regular in form that they can easily be 'grown' by computer. Fortey and Bell have produced such computer simulations by feeding in a few branching instructions for a 'given' proximal end and seeing the kind of colony that results. Just by permutating a few simple branching rules it is possible to 'grow' quite complex colonies, and many of these have their exact counterparts in nature. Presumably, then, the colony grew by a series of biological instructions comparable with those used in computer simulations.

What is striking is that the branched graptolites did *not* grow equally in three dimensions. Rather, most graptolites are an overall discoidal shape, or are conical (that is one can imagine the branches defining or embracing the sides of a cone). This shape has one important characteristic: if the colony moved upwards or downwards in the water column each zooid would have been able to feed clear of competition from zooids above or below. It is likely that graptolites were plankton feeders – or that they fed on organic detritus derived from the surface layers of the ocean. This would mean that an efficient system for harvesting food would have been a priority for the colony: they needed to make sure that the zooids within the colony did not interfere with one another in the capture of food, which may have been sparse at times. A planar or conical form is a consequence of this necessity. It is sometimes claimed that the colonies would have spiralled slowly as they moved through the water column, and if this were so, it would have given a zooid a greater 'sweep' through the water column than simple up-and-down movement without rotation.

It is possible to summarize the different kinds of many branched graptolites into a few categories, or branching strategies. Some of these branching strategies were adopted by different graptolites at different times in the geological record of the group. They represent solutions, as it were, to the problem of being planktonic, which were arrived at independently by different graptolite stocks.

1. *The deep cone.* This shape is characteristic of one of the first successful radiations in the group, exemplified by the genus *Rhabdinopora* in the Tremadoc (Fig. 41). This graptolite retains many primitive features from its benthic cousins, notably dissepiments, short black rods which link the numerous branches of many species, but there may be no reason to suppose that it was other than highly

specialised for planktonic existence. Its conical form may even have been derived from a planar ancestor as the geological record suggests. After the Tremadoc, the deep cone is rather rare except in the benthos, although it would be possible to regard *Monograptus turriculatus* (Fig. 35) as having re-invented this technique in the Silurian – but this time with but a single stipe defining the 'cone'.

2. *The dichotomising pattern* (Figs 82, 83 and 84). This is probably the commonest pattern, in which a spreading colony was achieved by simple forking (dichotomising) of the branches at more or less regular intervals. This kind of colony appeared at or near the beginning of planktonic graptolites. For the colony to grow without the branches interfering with one another a few rules have to be observed. The angle of dichotomy should decrease away from the proximal end; and a larger colony can be achieved by progressively lengthening the stipes from one dichotomy to the next. Graptolites really behaved like this (Figs 81 and 82). The result is a regular, discoidal colony. Dichotomising patterns can produce several variations on the basic pattern. In early graptolites the branches were produced by a complex budding sequence which mimics the structure of the first few thecae of the colony. Some later graptolites produced colonies which resemble some of these early graptolites, but they do so by means of cladia.

3. *Lateral branching*. Some of the most beautiful graptolites were produced by species which throw off lateral branches from two or more 'main stipes'. In some of these (e.g. *Goniograptus*; Fig. 117), the angles at which side branches are directed are controlled with quite extraordinary precision. Any change in angle from the one actually found would either produce interference between adjacent branches, or would produce a shape which did not effectively 'cover' the area embraced by the colony with feeding zooids. The shape actually found is both the most efficient, and the only one possible given the rules of branching observed by the colony. A variation on this is has been described as the Ying/Yang pattern (*Sinodiversograptus*, Fig. 57) from its resemblance to the oriental symbol. In this case, a main stipe is bent into an S-shape, but the lateral branches are splayed outwards in such a way as to cover the area enclosed by the whole colony.

4. *Spiral patterns*. This branching technique was adopted by Silurian monograptids (*Cyrtograptus*; Fig. 58) in which cladial branches are thrown off a central spiral; the branches may themselves branch further. Again a remarkable regularity in the production of branches – and even in the sequence of thecal changes along a branch – was the rule for such colonies.

What all these designs share is an efficient coverage of the overall space (discoidal or conical) covered by the colony: there are few 'holes'. This makes sense for an animal colony which is harvesting plankton or filtering organic material. After all, we do not expect to get a good return from a butterfly net or a trawl net peppered with large holes! So the graptolite colony presented an efficient 'net' to trap its planktonic food, while the design of the colony as a whole ensured that individual zooids did not have to compete with one another for the same food particle. The colony benefitted. It is scarcely possible to design such a cost effective harvesting array without making the design regular. There is the additional fact that the graptolite colony apparently needed to be hydrostatically stable as well – exceedingly lop-sided colonies would have twisted and turned through the water. All these features combine to explain the exquisite design of many-stiped graptolite colonies.

It is amusing to consider whether there are other possible designs for an efficient harvesting array, which have *not* been used by the graptolites. It is possible to make such designs for example by employing a maze-like motif. However, we do not know any graptolites with this shape – and why not? These shapes all involve right-angled 'bends' along a single stipe, and this was not possible for a 'real' graptolite: right angles could only be produced at a branch. Graptolite stipes can curve – as they do in the Ying/Yang and spiral patterns – but they very rarely kink (*Monograptus limatulus* and some other monograptids). So there were limits on the range of designs for efficient harvesting available to the graptolites simply because of the range of options for branching available in the construction of colonies. Bearing this in mind, it is in fact difficult to think of any design with efficient coverage which has *not* been used by one graptolite or another. I have managed to dream up one such which combines dichotomy and curvature. Perhaps this graptolite does exist, for example there is a *Dicellograptus* with *each* stipe individually coiled in a helical spiral, or perhaps others remain to be discovered. Or perhaps, after all, there is some constructional reason why such forms should be rare.

Graptolites with many branches were designed, therefore, as efficient harvesters in the open oceanic environment, and this explains why they employed such regular designs. It is perhaps not so surprising to find that other kinds of organisms that fed on plankton used similar shapes of nets. For example, certain crinoid arm feeding apparatuses are similar to *Goniograptus* in construction. What is more difficult to say is exactly how the graptolite fed. Did the zooids actively co-operate to produce a rise (or fall) through the water column during feeding? Or did they rise and fall in some diurnal rhythm in the manner of much of the living plankton? These kinds of questions are very difficult to attack from consideration of the design alone. It has been claimed that some graptolites were specially adapted to feed in the deep oxygen-poor sulfidic zone in the ocean. Certainly it is true that the many-stiped graptolites are commonest in very black anoxic shales, but also it could be claimed that their rather delicate colonies had a better chance of being preserved in this kind of environment than elsewhere. There is probably enough geological evidence to indicate that, in general, many-stiped graptolites were more characteristic of oceanic environments than their few-stiped relatives. In the early Ordovician, for example, one finds simple graptolites belonging to genera like *Didymograptus* and *Azygograptus* (Fig. 119) in relatively inshore sites, where they may be found interbedded with trilobite-bearing sediments with shallow water faunas. To find the complex branched graptolites such as *Sigmagraptus* and *Goniograptus* it is necessary to move to sites nearer to the Ordovician oceans. There seems to be no preservational reason why these genera should not be found with their simpler relatives, which leads us to conclude that there was a genuine ecological control on their occurrence.

It is probably reasonable to suggest that the more delicate of the many stiped graptolites lived below the turbulent surface zone of the ocean. Even though they may have been tougher than they looked, it is unlikely that the most slender forms (e.g. *Yushanograptus*) were as robust as diplograptids, or other genera with a back-to-back arrangement of thecae. These latter are often the only graptolites to be recovered from inshore sandstones, and they may have been typical of shallow epicontinental seas, and extended over the open oceans, too, in the shallower water layers. There may have been less competition for food in such shallower layers which meant that the spreadeagled, net-like system of plankton harvesting typical of the multistipes was not necessary for such species.

Chapter Six

WHERE AND WHEN DID THEY LIVE?

(a) THE WORLD THEN: 500 000 000 TO 400 000 000 YEARS B.C.

We suppose that the solar system was in the past much the same as it is now. The earth orbited the sun, spinning gyroscopically on its axis. Night followed day, and the earth's magnetic field was in operation and functioning satisfactorily. Velikovsky, by the way, was wrong: the moon was in place, and making its presence felt. We can recognise that sediments laid down in shallow water were stirred up and laid down in a manner that is quite characteristic of the effect of lunar tides. And more – the workings of the then celestial mechanics can – in some circumstances – be worked out from the particular patterns of preserved tidal sequences. Thus: the year used to consist of more (c.400) and shorter days, and there were about 13 lunar months in the year.

The earth has always had seas and continents, and the 'great' continents such as Africa have nearly always been dry land. The relative positions of the continents have, though, been wildly different in the past. A chart of the Silurian seas and coasts would greatly perplex a present-day mariner. The continental masses have wandered over the surface of the planet, driven by convection forces within the planet. Preserved magnetic particles within the rocks still point towards the magnetic pole of those distant days, and we can reconstruct the geography – to a degree – by making the magnetism of all of the rocks of one period point in the same direction. But beware of those palaeogeographic maps: latitude can be worked out in this way but not longitude!

The atmosphere was probably roughly the same as today's. There is no direct evidence for this in the form of, say, trapped air bubbles. Rather it is surmised on what did or didn't seem to happen then. Oxygen in the air reddened the land surface and enabled animals to breathe. Questions of CO_2 and CH_4 content are harder to resolve, being tied in part to the tricky questions of climate.

This is such a fiendishly complicated subject that one hesitates to broach it. Nevertheless, as graptolithologists tend to rush in where more sober minded geoscientists disdain to tread, . . . We suppose that the forces that drove the climatic engine were the same as today. But because the geography was different, the particular pattern of inter-related winds and ocean currents would have been grossly different from today's. Overall, the world's oceans did not boil nor, for the most part, did they freeze. This consistency may seem simple enough, but it isn't. Consider such features as a significantly lower output of energy from the sun at that time; a radically underdeveloped terrestrial flora and hence very different pathways in the CO_2 cycle; and the tendency of the present-day climate to react decidedly skittishly to relatively minor changes in incoming solar radiation, atmospheric composition and oceanic circulation. All this indicates that there have been some rather profound and highly effective 'mechanisms', regulating the earth's surface temperature for a very, very, long time. The 'Gaia' hypothesis of James Lovelock states that it is the totality of life on earth, functioning in concert as a kind of super-organism, which does this regulating. This idea has much aesthetic appeal, but be warned before raising it in staid scientific company. It has rather more detractors than supporters at present, and the former may splutter.

The seaways of the past are the great debating points of past geographical reconstructions. Almost all of the evidence we have derives from marine deposits, and yet paradoxically the most controversial discussions are about the opening and closing of ancient oceans. The physics of ocean currents were the same then as now, but the different geography of the day makes reconstructions of oceanic circulation patterns – and even of their specific motive forces – infuriatingly difficult to reconstruct.

The chemical composition is likely to have been roughly similar to the present, though it could well have been less polluted. Notions that the sea has been getting saltier through geological time, as more and more salts are washed out of the rocks, have tended to be replaced by the idea of a sea of roughly constant salinity (with salt being taken out of the sea as evaporites (salt deposits) roughly as fast as they are being put in). Casts of salt crystals have been found in rocks laid down before the graptolites were around.

We know quite a lot about the amount of dissolved oxygen in the seas of the time. At present, most of the world's seas are aerated: cold water pouring down from the polar icecaps carries oxygen to the very equator. Then, things might have been different. In shallow water, all sorts of animals and plants thrived in the well-oxygenated conditions. In deeper waters, 'events' of stagnation can be recognised in the sediments being laid down, with all traces of living activity, such as burrowing in the mud, being suppressed. The causes of these events are more obscure – were they worldwide in effect, relating to changed patterns of oceanic circulation, or were they local phenomena? We're still trying to work that one out. But such anoxic events were one of the prime causes of graptolites having a good fossil record.

(b) THE GRAPTOLITES IN THIS WORLD

Some species of graptolite are found from sites all over the world. Others are restricted to smaller regions. The study of this variable distribution depends for its accuracy on three main things: accurate geographic reconstructions of the period, precise correlation between faunas of the same age from all over the world, and an understanding of the factors which control plankton.

At certain times in the past the degree of restriction or localisation of graptolite species (or provincialism) was great; at other times it was much less pronounced and their geographic distribution was much more widespread. This localisation was strongly developed in the first evolutionary radiation of planktonic dendroids, which seem to have been divided into two rough provinces. The first of these includes areas of England and Wales, Southern and Central Europe, and is known as the Atlantic Province. The second includes parts of North America, Scotland, Australia, Scandinavia and Spitsbergen and is known as the Pacific Province. During the Lower Palaeozoic, these areas of deposition which made up the Atlantic Province were arrayed on either side and some distance from the Equator. Species like *Rhabdinopora flabelliformis* were confined to the Atlantic, or high latitude, area while other dendroids and anisograptids were common in the Pacific or equatorial region. It may be that planktonic forms developed independently in several places, accounting for this early provincialism. Certainly in the late Tremadoc provincialism waned considerably with most genera having a wide distribution, even rare forms like *Psigraptus* (Fig. 121).

The Atlantic and Pacific provinces were re-established in the Arenig by the new graptoloid plankton. This is possibly the most dramatic phase of provincialism (David Skevington 1974). At this time many genera are completely restricted to rocks of the Pacific Province. The list is a long one – *Brachiograptus, Cardiograptus, Paraglossograptus, Pseudobryograptus, Allograptus, Sinograptus, Holmograptus* and others. This was clearly an area of considerable diversity and speciation. By contrast the Atlantic province has a much lower diversity of graptoloids. Although some species and genera, e.g. *Azygograptus*, are limited to it, this province is most easily recognized by the absence of so many Pacific forms.

The high diversity Pacific province is also the equatorial one and this fits well with patterns in the distribution of modern zooplankton. In general the number of species increases today towards the Equator while the abundance of a single species decreases. Species in higher latitudes today seem to be generally larger than their relatives in the tropics. The similarity of diversity and size patterns indicates that the Arenig provinces were results of latitudinal differences rather than of any other change, although it has been suggested that they are depth related assemblages, with the Atlantic Province being a deepwater area and the Pacific a shallow one (Berndt Erdtmann, 1976). However, this hypothesis seems to fit the overall pattern less well.

Although these provinces are clearly defined, they are by no means mutually exclusive. Many areas yield a mixed fauna, especially those which would have lain at intermediate latitudes like China and South America. Even within one province, there are many differences in the faunas from different places. For example, both Spitsbergen and the Bendigo region of Australia lie within the Pacific

Province. Of the thirty or so species at both places, only seventeen are common to both (Barrie Rickards and Amanda Chapman *in press*). This is not surprising when the size of the provinces is considered. The North Sea and the coastal waters off Newfoundland today are both at similar latitudes, yet the sea off Newfoundland freezes in winter and icebergs are common even in the summer months. The fauna of both has many similar components, but no-one would expect them to be identical given this radical difference in temperature. The difference is due to current systems in the ocean. Off Newfoundland the water comes straight from the Arctic as part of the Labrador current. On the other hand the North Sea is warmed by water of the North Atlantic Drift which originates in the Caribbean. Current control *within* major graptolite provinces was suggested by Bill Berry as long ago as 1960.

What other factors might have been responsible for controlling where certain graptolites could live? Latitude differences work mainly through changes in surface temperature of the sea, but as we have seen, this is also affected by currents. Temperature also changes with depth as bottom waters at present tend to originate at high latitudes and flow towards the Equator. Depth has long been considered a possible control on graptolites. Apart from the suggestion that the Pacific and Atlantic provinces are depth-related, a convincing argument has been put forward for isograptids being deep water species. Detailed work by Richard Fortey shows that genera like *Isograptus* and *Pseudisograptus* (Figs 18 and 51) occur around many continental masses well offshore and associated with deep water species of benthos. Different assemblages occur in shallow water.

In modern plankton, one of the most profound barriers to distribution is found near the edge of the continental shelf. The near shore, neritic zone, is shallow and can be affected by local fluctuations in 'climate' like the output of rivers or cyclones and is a generally stressful environment for life. The deeper area seaward of this has a much more uniform 'climate' and species tend to be much more widespread. Surprisingly, even plankton which lives in near surface water changes at this boundary; it is not just plankton requiring deep water which is lost. It is possible that a number of examples of neritic graptolite faunas have already been documented. One was found by Finney (1984) in Eastern North America. Another has been found in rocks of Silurian age (Barrie Rickards, Susan Rigby and Jon Harris, 1990). It seems to those authors that *Pristiograptus tumescens* is more or less restricted to shelf environments, whilst *Saetograptus incipiens* is an offshore species at the same time. *Azygograptus* (Fig. 119) distribution in the Ordovician affords another example. After the Arenig and Llanvirn periods, provincialism declined and during the whole of the Silurian it is difficult to detect. This may be due to changes in global climate or changes in sea level – there is a well-documented Ice Age at the end of the Ordovician – or to evolution of better adapted graptolites which out-competed previous forms in both provinces. Whatever the cause, provincialism at generic level was rare to absent in the Silurian. However, at species level it was still present. During the Wenlock three subprovinces have been recently recognised. One of these seems to be the remains of the old Pacific tropical province. A second in the Mediterranean is a high latitude fauna. The third occurs in the Baltic area which was then at low latitudes. This Rheic subprovince was probably situated over a shallow water area and may represent an example of a neritic assemblage (Barrie Rickards, Susan Rigby and Jon Harris, *op. cit.*).

In all the preceding examples of graptolite provinces and assemblages there is one underlying requirement – and that is a thorough understanding of which graptolites coexisted. This is simple in terms of largely cosmopolitan faunas, because there is a good chance of different graptolites occurring in the same locality. In times of high provincialism the problem is much more acute. A striking example of this comes from the correlation of Arenig and Llanvirn age assemblages from Britain and North America. Both areas apparently had a *'bifidus'* zone characterised by the presence of *Didymograptus bifidus*. North America also had a zone characterised by the presence of *Paraglossograptus etheridgei*. In 1973, Skevington postulated that the *'bifidus'* and *'etheridgei'* zones of Britain and North America respectively were coeval. In 1982 this was finally confirmed by Cooper and Fortey who showed that the two *'bifidus'* were actually different species which had rather different origins. Had the reverse been true, then many provincial assumptions would have been groundless.

The other necessity for reconstructing graptolite provinces is an accurate reconstruction of the geography of the world in which they lived. Palaeozoic reconstructions are being constantly improved, and many apparent anomalies are disappearing as this happens. However, it will be a long time before we can be certain about where any more than a minority of the graptolites lived, and why they lived there.

Chapter Seven

HOW COMMON WERE THEY?

There is little point in trying to claim that graptolites are very common to a professional geologist who has just told you to your face that in his whole career he's never managed to collect one! And this despite looking for them on field trips at localities where they were supposed to occur, and where they had proved crucial in some stratigraphic debate. Cynicism grows rapidly under such circumstances. Nevertheless graptolites are much more abundant in Lower Palaeozoic marine strata than frequently supposed, and on the photographs in this book you can see slabs covered in them. Of course, as we emphasize in several places, some groups such as dendroids are just rare as fossils especially those benthic forms which lived in the inshore shell-rich and clastic environments: occasionally dendroids may be locally abundant in such facies. Planktonic graptolites may also be sparsely distributed, and of low species diversity, in the same environments, even though they may be very important in the correlation of such facies into the offshore environments.

Our professional geologist referred to above should have been aware of these particular constraints on graptolite abundance: his complaint is against so called graptolitic shales or slates! Chs 2 and 4 give some guidance as to what to look for in the field to avoid disappointment, so I shall not repeat those points here, but rather investigate the actual numbers one might expect to collect, and the actual numbers of graptolites preserved and perhaps potentially collectable.

If I can begin with some actual examples: two of the authors of this book, whilst working for their PhDs in the Lake District, collected over 20,000 Silurian graptolites each. (These collections are now in the Sedgwick Museum at Cambridge). On a separate, later, occasion I was informed that collecting Wenlock graptolites in the Lake District was a waste of time, my informant having seen only five specimens in several years of work there. Three days after beginning work two very large wooden crates had to be made, total volume one cubic meter, to hold my collections! The simple explanation of this discrepancy was that the failed collector was a sedimentologist who looked at the cleavage and joint faces of the rocks, in order to examine the sequences of sedimentary rock types in section, whereas I split all the rocks on the bedding plane having first found the rock types described in Chapter 2.

On the subject of rock type I will deal with only one aspect not discussed in

Chapter 2, namely the clue given by the presence of laminated, dark, carbonaceous matter – the so-called hemipelagites. Rocks which have this characteristic lamination whether they are black shales, bluish mudstones, brown siltstones, sandstones or limestones are likely to yield large numbers of graptolites, for the carbonaceous matter comes from the same planktonic source as the graptolites. Every experienced field collector knows how to spot this clue – a texture rather than a specific rock type – and homes in on it quickly.

Once happily splitting the rock along the bedding plane we can turn to the question which forms the title of the Chapter, namely just how common are they? In dark graptolite shale, where condensed deposition is the rule, then a great abundance is likely, and great diversity on a single bedding plane is probable (Fig. 123). Examples are the Hartfell and Glenkiln Shales of the Southern Uplands, the Skelgill Beds of the Lake District, or the black shales of Abereiddi Bay in Pembrokeshire. I have made counts of the numbers of specimens per unit area in both the Skelgill Beds and in the Utica Shale of eastern North America, and the results are rather surprising, as I shall relate below.

However, in sequences where the sedimentation is more 'spread', as in turbidite sucessions, where graptolitic beds are often common, the laminated shales commonly exhibit few specimens and a much lower species diversity upon individual bedding planes. Many bedding planes will have but a single species, although all growth stages of that species may well be present, from individual prosiculae to the adult rhabdosome some several centimeters long. While there are certain problems in explaining their occurrences, a first approximation might be that they record individual plankton kills of monospecific plankton clouds from particular niches. Very little study has been made so far of such individual bedding plane 'populations' but they might well provide useful insights into aspects of graptolite ecology such as population structure (see Douglas Palmer, 1986).

In the condensed black shale sequences it is possible that we are seeing, therefore, a succession of such plankton kills that arrived on the substrate where very slow deposition was taking place, with many minute non-sequences (so far as sediment is concerned). That the graptolites in such circumstances did arrive at different times is easily established by the differing presentational modes on the one bedding plane.

It follows that in order to attempt a calculation of the number of individuals in a plankton cloud, then the 'spread' sequences are likely to be best; but to calculate the number of individuals preserved in the classic black graptolitic shale one must avoid figures that come from 'spread' sequences.

In the Utica Shale and the Skelgill Beds I found the numbers very similar and will quote here only the latter. An underestimate would be 10 graptolites per square foot; and the areal extent of the Skelgill Beds is 200 square miles. Thus the number of *preserved* specimens on one bedding plane would be about 30 million. The thickness of the Skelgill Beds is perhaps 50 feet; and an underestimate of the number of bedding planes per foot might be 25. Thus the number of preserved graptolites in the Skelgill Beds is around 40,000 million. All these figures are gross underestimates in one sense, in that they refer to preserved specimens, those that survived the taphonomic processes. In another sense they overestimate the numbers, because a high proportion of graptolites, even in the black shale environments, are broken, presumably after death and before being buried (I say after death because those broken during life are known to regenerate). Peter Crowther did this work during his own PhD studies, and he found that on some surfaces over 80 per cent of rhabdosomes were broken. (That they were infrequently bent confirms their relatively rigidity during life and after death). So it is possible that counts may record broken specimens more than once.

Whilst it is possible to accurately record the numbers of specimens or part of specimens per unit area, the figures per unit volume can only be underestimates, simply because there seem at times to be an almost infinite number of bedding planes, say in the Skelgill Beds black shales, with abundant graptolites.

Although I made the point above that the rock must be split parallel to the bedding plane to collect graptolites, the experienced collector can see them in surfaces broken at a high angle to the bedding (Fig. 40 is a somewhat extreme example, of 'sandstone' grade). Thus some idea of the number of graptolitic layers per unit thickness can be (under) estimated: a rough guide that this statement is correct can be obtained by dissolving a piece of graptolite-bearing limestone (Chs 2 and 11) on which no specimens are visible, and counting the subsequent yield in tens of specimens if not hundreds.

It follows from all the above that graptolites are actually quite common in the graptolitic facies whether this is black graptolitic shale or siltstone layers amongst turbidite sequences. What is needed to realize this fact is a measure of persistence, it is true; but more importantly the ability to recognize the rock type and work it correctly.

CHAPTER EIGHT

WHAT OTHER ORGANISMS DID THEY LIVE WITH?

This apparently straightforward question is surprisingly difficult to answer. For most graptoloids there is no direct evidence of what other organisms they lived with. Perhaps 'associated with' is a better term since there is remarkably little interdependence between marine invertebrates outside reef communities, largely because the food chains are so short. The main difficulty is caused by our inability to observe the ancient planktonic life habit directly so that we depend on indirect evidence from post-mortem accumulations of skeletal remains in the sediment. Even the life habits and associates of the benthic dendroids and other minor orders of graptolites are still not well established and understood.

Very few rigourous studies have been made of graptolites in relation to their associated faunas. One of the main reasons for this may be simply that graptolites are not commonly found in the company of other fossils. For instance in black shales (such as those at Abereiddi) graptolites can occur in densities of thousands per square metre of bedding plane with hardly any other fossils being observed. Conversely, shell bearing sediments generally do not contain graptolites. A recent detailed survey of Ludlow (Silurian) shelf sediments and faunas identified 33,000 shelly fossils but only a hundred or so graptolites (Watkins and Berry 1977). This dichotomy reflects a long held distinction between what has been called the 'graptolite shale facies fauna' and the 'shelly facies fauna'.

Why should there be such a distinction when graptoloids are generally considered to have been predominantly planktonic? As plankton they could be expected reasonably to have had a very widespread distribution within the marine realm and have been carried by wind and tide into shelf sea areas, potentially to breed there and eventually be preserved within the sediments there. Equally the dendroids and other benthic graptolites could be expected to be found as frequent associates of the common shelly faunas within the deposits of the Lower Palaeozoic shelf seas but are not.

There are two main possibilities. Either graptolites did originally occur in shelf-seas but their remains were destroyed before being fossilized or, their associations were largely excluded from these areas.

Firstly the question of preservation potential. We have seen (Ch. 2) that the organic material of the graptolite skeleton is remarkably resistent to chemical attack. Therefore it is unlikely that their skeletons would have been removed from within sediment by post-depositional solution as commonly occurs to carbonate shells. Physically the skeleton seems to have been quite tough and flexible and resistant to the sort of hydrodynamic forces encountered by organisms of their size within the sea. However, rhabdosomes are likely to have been destroyed by contact with coarse sediment in high energy environments such as the tidal surf zone.

We do know that they could survive some injuries and even breakage because rhabdosomes have been found with clear evidence of repair and regeneration (Fig. 126). What we do not know is whether the damage was caused by biological or physical agents but the former is more likely. So in general we should not expect to find graptolites or, at least those graptolites with particularly delicate skeletons, in the deposits of high energy environments. However, as we shall see below there can be exceptions.

There are many other quieter water shelf sea environments where low energy conditions prevail and fine grained sediments are deposited. Why are graptolites not found here? Well, sometimes they are, but not generally when the fauna is dominated by the common shelly invertebrates of the Lower Palaeozoic such as articulate brachiopods, trilobites, corals, bryozoans, etc. Whenever these relatively shallow water shelf environments (< 100 m deep) are well oxygenated they can support high density and high diversity faunas including many soft bodied infaunal burrowers. The presence of these organisms without mineralised skeletons is recorded by traces of their activity remaining in the sediment after lithification. The resulting biological disturbance (bioturbation) of the sediment can be very intense and have an effect similar to that of earthworms in soil. Primary sedimentary structures are destroyed, organic detritus is removed from the sediment and possibly organic walled fossils such as graptolites.

In conclusion it would seem that there probably has been a considerable post-mortem destruction of graptolite remains by both physical and biological means from within particular shelf environments. But, there is still a lot that we do not understand about these processes of burial and preservation (the taphonomy) of graptolites.

There is another factor which needs to be taken into account and that is the distribution of plankton in relation to the major water bodies. This has already been referred to in the discussion of palaeogeography. It has been shown that at the present there are relatively well-defined boundaries between oceanic and shelf sea water bodies which roughly coincide with the edge of the continental shelf. Furthermore, distinctive plankton assemblages characterise these separate waters and their skeletal remains are separately recruited to the sediments underlying each water body. Consequently their fossil distribution can be used to distinguish the original parent water body. If this sort of biogeographic and hydrographic distinction applied in the past then we should expect to find differences between the plankton assemblages and associations of shelf seas and open oceans. And, we should not expect the open ocean plankton to encroach right across the shelf seas and be found fossilized there in anything like the same sort of diversity or abundance.

It has also been argued that different graptoloids lived at different depths. If this were true then only those that occupied the upper 150 metres or so would have been able to enter shelf waters. These hypotheses of graptolite distribution have yet to be thoroughly tested.

So if we conclude that the majority of graptoloids were not primarily part of the shallow shelf sea plankton, what were the non-benthic inhabitants of the Lower Palaeozoic pelagic environments before the evolution of fish with jaws armed with teeth? The graptolites must have fed on something and in turn have been the food supply for others; no organisms exist in isolation outside the food chain.

The primary producers of the Lower Palaeozoic, those microscopic unicellular plants of the phytoplankton, include a heterogeneous group generally known as acritarchs. Their tiny (5–500 μm) organic walled skeletons are preservable and they can be very common as microfossils in graptolite bearing sediments. There were also a few groups of preservable microscopic primary consumers in the zooplankton quite apart from the myriads of non-fossilisable larvae of the marine benthic invertebrates.

A window into this ancient world of microplankton is provided by some very impressive sequences exposed in the various islands of the Canadian Arctic Archipelago. Here shales and limestones of Ordovician to Devonian age are often highly fossilferous. They are considered to be the deep water equivalents of shallow shelf carbonates and as a result contain mainly pelagic faunas that can be (acid) etched from the limestones. The beautifully preserved, unflattened fossils that have been isolated from these rocks include two groups of primary consumers, the organic walled tests of an extinct group called chitinozoans (5–200 μm) and exquisite spiny skeletons of radiolarians (20–200 μm) made of silica, a group that are still important members of the zooplankton today. The fauna includes the previously mentioned phytoplanktonic acritarchs, conodonts (see below), graptoloids and abundant siliceous sponge spicules. These sponges are vase shaped sedimentary filter feeders and are about the only skeletised benthic organisms that were, and still are adapted to these relatively deep basin environments.

It is likely that the planktonic graptolites with their numerous zooids in each colony were also primary consumers drifting with the multitudes of plankton and feeding on it. Present day analogues would perhaps be cnidarians such as the smaller 'jelly-fish' and the ctenophores ('comb-jellies' or 'sea-gooseberries').

Black graptolitic shales often contain shapeless and generally indeterminate carbon films scattered on bedding planes with the graptolites. It was this close association with degraded 'sea-weed' like material that led Charles Lapworth, a hundred years ago now, to make a comparison with the present day organic masses of floating *Sargassum* weed. By this analogy he argued that graptoloids could have been originally attached floating algae by their nemae (i.e. epi-planktonic). He considered that, periodically, parts of the mats broke off and foundered to the sea bed taking their graptoloid and associated epifauna with them.

Although it is possible that some graptolites may have had this habit, it is now considered that they were independently buoyant. Nevertheless, as we shall see, the *Sargassum* analogue has been maintained in an attempt to explain the occurrence of some shelly invertebrates found in association with graptoloids.

By far the most common macroscopic fossils associated with graptoloids are the simple, elongate, calcium carbonate cones of cephalopods. They are generally straight and narrow conical shells of up to 200 cm or more in length and 5 cm or so wide at the aperture when flattened. These orthoconic nautiloids, as they are called, housed a squid-like tentacled animal within the apertural body chamber. They are generally regarded as having been active swimmers and predators with a shoaling habit. Thus they probably filled some of the niches now so successfully filled and dominated by some of the bony fish, whilst the living distant descendants of orthoconic nautiloids the squids and cuttlefish have a more secondary role.

It is not known what these ancient cephalopods fed on. However, as with the graptolites, on death their skeletal remains sank to the sea bed and were fossilised in the sediments accumulating there, often in great numbers. When abundant they form cephalopod limestones or coquinas with graptolites and very little else. The fossils are often orientated with their long axes subparallel to one another. Such alignments indicate the presence of bottom currents and measurement of their flow directions is important in the reconstruction of the original submarine topography.

The remains of other groups of swimming (nektonic) animals, mainly arthropods of different kinds are also occasionally found with the planktonic graptolites. They include a group of crustaceans called phyllocarids. Quite large (up to 7.5 cm long) and vaguely prawn like swimmers they had a laterally flattened, bivalved, thin carapace that completely covered the head. They had powerful mandibles but in the absence of grasping appendages it is unlikely that they were predators, rather scavengers feeding on carrion.

Also, there are rare fossils of some extraordinary extinct merostomes, the eurypterids. Elongate, dorso-ventrally flattened scorpion-like creatures, some of them were two metres long but mostly they were about 40 cm or so. They had a pair of appendages developed into swimming paddles that operated in conjunction with a flat flange shaped tail. These animals were active predators and possibly cannibalistic with grasping spined anterior appendages and strong mandibles. As a group they were also extremely adaptable ranging through most marine environments and into freshwaters at the end of the Silurian.

During the Silurian too there was another group of small crustaceans the ostracodes, most of which were and still are benthic but they include some forms that invaded the plankton at this time and still have representatives living there. These are the extant myodocopids and they have only partly mineralised bivalved carapaces (up to 2 cm in diameter) that virtually enclose the animal except for an anterior opening from where the swimming appendages extend. Typical of many planktonic organisms they tend to have considerable geographic ranges. For example some of the Silurian ones have been found in marine sediments from Britain through Southern Europe to Australia.

Trilobites are perhaps the most famous extinct group of fossil invertebrates and were, in taxonomic diversity, the dominant group of animals in the Lower Palaeozoic seas. However, these arthropods were for the most part members of the shelly shelf benthos living amongst the brachiopods and corals. A minority of small (2–5 cm long) and often spinose forms such as the rhaphiophorids are occasionally found with the planktonic graptolites in the Ordovician (but not in the Silurian). It is thought they they were probably swimming forms that fed on plankton and other small particles of organic detritus.

Another enigmatic extinct group of organisms that should be mentioned in this context is the Conodonta. Tiny serrated tooth-like elements only a millimetre or so in size have been etched in large numbers from shelf limestones for over a hundred years now. It is only in the last few years that any real understanding of their biology has been achieved. For the first time some conodonts were found with clear traces of their related soft tissues. It appears that they were small, elongate (4 cm or so) and narrow, laterally flattened worm like creatures with assemblages of the elements clustered anteriorly as an intermeshing apparatus for feeding. Superficially they look like the living chaetognaths (arrow worms) but they have features indicating that they may have been jawless craniates. Although most conodonts were associated with shelf seas, there are also deeper water low diversity associations, with graptolites, that have widespread distributions and clearly occupied the pelagic realm as active swimmers.

There is a famous and protected Site of Scientific Interest (SSSI) in the south of Scotland near Moffat called Dob's Linn. Here Charles Lapworth worked out the ordered stratigraphic change in graptolite assemblages through a sequence of black and grey shales. His publication (1878) established Dob's Linn as the standard reference for the Moffat Shale and proved beyond all doubt that graptolites often provided the only practicable means of correlating Lower Palaeozoic strata.

Just over a hundred years later this same section has been internationally recognised (by the International Union of Geological Sciences in 1985) as the 'boundary stratotype' for the junction between the Ordovician and Silurian Systems. Because of its importance for global correlation a great deal of attention has recently been paid to the strata of this locality and their faunal potential.

Graptolites are the dominant fossils of these deep water black shales along with the microfossils discussed above, the acritarchs and chitinozoans. Accompanying them are low diversity conodont assemblages typifying off shelf waters and in addition rare scolecodonts. These latter are the preservable organic biting jaws of polychaetes (bristleworms), a diverse group of still living, largely aquatic annelids. Superficially they resemble conodonts but differ in their composition and microstructure. It would seem by their rare occurrence in the graptolite facies that some of them, like conodonts, were nektonic swimmmers.

One major group of invertebrates which has barely any planktonic representatives as adults is the Phylum Echinodermata. However, in the late Silurian some crinoids (sea-lilies) evolved a bulbous, tuber like buoyant structure that floated them upside-down. Consequently these, the scyphocrinitids, had a very wide distribution that included the domain of the graptolites.

Planktonic graptolites are also at times found associated with the shells of various invertebrate groups that are normally common members of the shelf sea shelly benthos (Fig. 125), especially certain brachiopods and bivalves. These shells may be quite abundant and are generally distinguished by their small size (less than 15 mm) and thin shells. The brachiopods belong to various lingulid and strophomenid genera. The lingulids are normally semi-infaunal with long fleshy stalks that protrude between the shells and anchor them within burrows in the sediment. The strophomenids were normally epibenthic with similar stalks for attachment but only as juveniles; their pedicles atrophied when they became free living adults.

Neither of these groups would seem to have been very good candidates for an epiplanktonic or planktonic mode of life. It has been suggested by Bulman, following a known phenomenon of modern ocean plankton, that the lingulids may have been 'giant larvae' derived as normal from neritic parents but retained in the open sea plankton by lack of suitable sites for settlement. The analogy might be extended to the strophomenids but modified to have them as epiplankton because of their relatively heavier carbonate shells compared with the chitino-phosphatic shells of the lingulids.

The same argument has been used to explain the presence of various small bivalves in the graptolite facies. Most of these bivalves belong to one of the very few major groups of bivalves to have become extinct, the praecardioids. In these the epiplanktonic attachment to floating algae (the *Sargassum* model again) is supposed to have been achieved by organic byssal threads similar to those of the modern common mussel (*Mytilus*).

However, all this is highly speculative and few of these brachiopods and bivalves have actually been found in close association with anything resembling floating algae. It is equally possible that they may just have been the few species that were adapted to an epibenthic existence in waters with low levels of dissolved oxygen (dysaerobic) or low nutrient levels.

There are other larger bivalves that are more common in some graptolite facies. Pterineid bivalves (1–2 cm) can be found in densities up to 400 per square metre with all growth stages and their shells still joined. The locally less common but geographically more widespread and larger shells (up to 5 cm) of the cardiolids tend to have similar growth characteristics and articulation. These features indicate that they originally lived at the same sites where they are found as fossils, that is they were genuinely epibenthic. The association again may best be explained by adaptation to low oxygen or nutrient levels. Certainly there was some stressful factor in the environment that excluded the normal shelf shelly benthos with the dominant brachiopods. So these are mixed assemblages where dead graptolites accumulated on the sea bed intermingling with the bivalves, the only organisms that could live there. The host sediments tend to be grey laminated muddy silts. This facies and its graptolite-bivalve-orthocone fauna are typically found fringing the shelf facies and are separate from the euxinic black shale facies.

The dendroid graptolites are for the most part sessile epibenthic organisms that grew up from substrate attachment of the sicula into the typical branched bushy, tree like or conical form. They are found generally in fine grained often carbonate rich shelf sea sediments. The distribution of most species is geographically restricted so they are endemic forms compared with the generally more cosmopolitan planktonic graptolites (graptoloids). Considering the overall duration of the dendroids as a group (some 150 Ma) and that they outlived the graptoloids by some 50 Ma, they clearly were a successful group. Nevertheless, they are also remarkably rare. There are professional palaeontologists who have spent their working lives looking at Lower Palaeozoic strata and fossils the world over without finding or seeing a single dendroid and yet they will have found many graptoloids.

The question of dendroid taphonomy in relation to the shelf sedimentary environments has never been investigated. It may well be that they were originally a lot more common but have been destroyed by bioturbation. Certainly fragmentary and unidentifiable dendroids are quite often recorded in fossil lists from a wide range of shelly facies through-out the Lower Palaeozoic. Also they would have been particularly prone to being uprooted or torn from their holdfasts by major storms in shallow waters and would not have been able to reattach themselves.

It is not surprising that our knowledge and understanding of these little colonies and their ecology is still quite primitive. Indeed it has even been claimed not so long ago (by Bouček, 1957) that they were epiplanktonic on floating algae (the old *Sargassum* hypothesis again) because some large dendroid colonies seem to be associated with no other fossils except graptoloids. Certainly the form of many of the rootlike holdfasts does not necessarily prove attachment to the sediment but then neither are the 'associated algae' found. It is another area that needs investigation.

There are only a few localities and horizons around the world where dendroids can be found quite readily. One of the best known is the Gosport channel lens between the Lockport reef limestones (Silurian) of New York State. The shales and sandstones of the channel contain dendroids, a few small inarticulate brachiopods, worms, and algae whilst the flanking reefs have rich shelly faunas with corals, crinoids, molluscs and brachiopods. Rare dendroids are also known from the flanking shales and mud drapes to reefs and carbonate mud mounds from the Llandovery of the Yangtse platform in South China to the Wenlock of the Wren's Nest, Dudley (England) and the Lower Carboniferous of Feltrim Hill Quarry, near Dublin (Ireland) and Pendle Hill in Yorkshire. Is it that the dendroids only lived in the quieter slightly deeper channel waters or is it only there that they are preserved?

In the Lower Palaeozoic strata of the Barrandian (in the Prague Basin, Czechoslovakia) dendroid faunas have been found at several different horizons. Again they tend to occur in fine grained sediments with restricted faunas of inarticulate brachiopods, graptoloids and rare cephalopods. Although the rhabdosomes are sometimes virtually complete it is very difficult to differentiate the details of the holdfast because they still retain some three dimensional form at an angle to the bedding planes.

A different sort of faunal association has been recorded from the Silurian of the Pentland Hills in Scotland. Again there are fine grained shallow water sediments but the diverse dendroid fauna is found with eurypterids, spiny echinoids, starfish and a curious calcareous alga (*Ischadites*). Some of the dendroids have fine root 'fibres' for anchoring the colonies in the soft sediment but there are a few others that have been found attached to brachiopod shells. This facies fauna is a rare one probably because it normally has a low preservation potential; nevertheless its existence supports the notion that the dendroids were more common than the preserved record shows.

There is quite a range of other epibenthic graptolites and graptolite related colonial organisms that probably occupied similar ecological niches to the dendroids but are rarer still. They include the tuboids (Figs 45 and 46), camaroids, stolonoids, crustoids and pterobranchs and are all very small (5 mm) encrusting forms about which there is only limited information. Many of them are only known from a single extraordinary locality in southern Poland.

In the 1930s Roman Kozlowski and a colleague were searching some Tremadoc (lowermost Ordovician) glauconitic sandstones and cherts for small chitino-phosphatic shelled inarticulate brachiopods. A few species of these were found in abundance in both the sandstones and cherts but only in the latter were they well preserved. They searched the cherts by flaking them into thin transluscent shards that could be examined with a handlens. Whilst doing this they noticed some other tiny (< 5 mm) organic walled fossils. By systematically working through some cubic meters of chert with this painstaking method they were able to select some 260 kilos of fossiliferous shards to take back to the lab in Warsaw. There the bulk was further reduced to just 18 kg for acid solution. Because the fossils were in chert they could only be dissolved by hydrofluoric acid, a highly dangerous and in those days very expensive acid (see Ch. 11 for details). These laborious methods took some years to complete but the reward was great, a veritable graptolitic bonanza.

Kozlowski had discovered a totally new graptolite fauna from which he was able to describe fifty new species and three new orders the tuboids, camaroids and stolonoids. There was the first known fossil representative of the living pterobranchs and forty further species of other graptolite related organisms. However the path to publication of his results was a uniquely fraught one (see biographical note, p. 000).

The associated fauna of this amazing diverse benthic graptolite microfauna consisted predominantly of the previously mentioned low diversity but highly abundant inarticulate brachiopods, numerous separate siliceous sponge spicules, a few fragments of tribolites, two conodont elements and very numerous microscopic chitinozoans but no carbonate shelled fossils.

In general terms this quite coarse grained clastic deposit originated in shallow waters nearshore with relatively high energy conditions, which is why the graptolites are such small early-growth stages and largely fragmentary. This sort of facies would normally be a most unpromising prospect for graptolites. The only reason the tiny fragments survived is because they were enclosed in a silica gel that hardened to form chert before the sediment was fully compacted. As a result very delicate thecal structures were kept intact and protected from subsequent deterioration for 500 Ma.

What is not clear from this assemblage is what the graptolites encrusted, but it may have been marine algae from which they became detached as it degraded.

Kozlowski subsequently extended his researches into this group of graptolites by etching a wide range of nodules and boulders of Ordovician to Silurian age. They were mainly glacially transported boulders (erratics) derived from Polish tills and moraines that originated as limestones in Scandinavia and the Baltic. They yielded further extraordinary examples of the graptolite microcosm with different combinations of the following fossils: graptoloids, dendroids, tuboids, crustoids, camaroids, pterobranchs (all hemichordates); inarticulate brachiopods, sponges, conodonts, scolecodonts, foraminifera, acritarchs, chitinozoans, hydroids and algae. These are of necessity all non calcareous organisms because the process of dissolving limestones also destroys calcium carbonate shells. However the assemblages do generally confirm the evidence of the bedding plane associations described above and suggest that the graptolite facies was a particular one dominated by remains from the plankton but mixed with a limited range of benthonic organisms.

CHAPTER NINE

WHAT WAS THEIR SEX LIFE LIKE?

Almost all graptolite remains, that most collectors usually find, are the asexually produced rhabdosomes or colonies (Ch. 1). However, as graptolites did not reproduce by jettisoning grown branches or stipes in order to become new individuals, there must have been another side to their sex life. There can be no question but that graptolite colonies reproduced sexually at intervals. In fact graptolite eggs are known: Roman Kozlowski (1949) detected them within encysted thecae of encrusted colonies; he further reported graptolite embryos, hatched within the encysted thecal cavity; and this last observation was confirmed by Oliver Bulman and Barrie Rickards (1966) working upon upright bushy tuboids (Fig. 1) from the Canadian Arctic. Despite this evidence, we know frustratingly little about these early developmental stages in the (sexual) reproductive cycle of graptolites.

The next graptolite stage seen is the prosicula (Fig. 13) and it seems likely that the extruded embryo rapidly secreted around itself the well known sharply pointed, conical sheath. The embryos within the thecae closely resemble the embryos described by Stebbing (1970) within the thecal tubes of modern rhabdopleurans: they too are extruded (through rather narrow zooidal apertures) and subsequently secrete a boot-shaped sicula, presumably very shortly after spat settlement occurred. Thereafter, in both graptolites and rhabdopleuran hemichordates, colony growth is by asexual budding producing the linear colonies illustrated in this book. But there are other implications of sexual reproduction, which we consider below.

Graptolites were genetically adventurous. The evolutionary dynamics resulting from the amazing morphological diversity of graptoloids in particular, is remarkable. Very little is known for certain about their reproductive habits. As a result, we can only speculate about what made graptolites such a successful Lower Palaeozoic group.

Graptolites' nearest living relatives, pterobranchs such as *Rhabdopleura* and *Cephalodiscus*, reproduce both sexually as well as asexually. The majority of zooids in colonies of *R. compacta* are either neuter or sexually immature; however, some individuals do mature sexually and contain either developing ova or a testis. The tentacles and digestive systems of these sexual zooids are degenerate, and presumably they must receive nutrition from the rest of the colony via the stolon system. Zooids of *Rhabdopleura* have only one gonad and are therefore not hermaphroditic but gonochroistic. Some individuals of *Cephalodiscus* have both ovary and testis and are thus truly hermaphroditic.

Similar gonochroistic differentiation of zooids probably occurred in dendroids. The female zooids are thought to have been housed in the large autothecae and the males in the small bithecae (Fig. 16). Dendroids outlived their graptoloid successors, although they never showed the same evolutionary flair for diversity.

Compared with this sexual conservatism of the dendroids a dramatic change occurred in the planktonic graptoloids. The small bithecae were lost, and the rhabdosome had only one type of theca, in structure presumably equivalent to the autotheca. This probably points to a change in sexuality from gonochroism to hermaphroditism. At first sight the hermaphroditic habit would seem to be a disadvantage. All zooids in a graptolite colony contain the same genotype. Prolonged self-fertilization would surely eventually weaken the stock. However, hermaphrodites in the animal kingdom have generally developed stratagems to avoid self-fertilization. Graptoloids, apparently, revelled in their hermaphroditic state; having acquired it, they proceeded to undergo an evolutionary explosion!

How did graptoloids avoid in-breeding? There does not appear to be any evidence, as in Bryozoa, for the cyclical control of sexuality by the colony. By analogy with other modern hermaphrodites, about which there is relative ignorance, graptoloid zooids were probably bisexual for only a short period of time. Depending on its developmental stage, a zooid could have had the potential to be a temporary male, female or neuter. Zooids maturing at the proximal end of a graptoloid rhabdosome might initially have been males whereas those occupying the larger distal thecae were more likely to have manifested the female phase. Those zooids budding asexually from the growing end(s) would have been immature. It is tempting to speculate further that in biform graptoloid colonies, a permanent sexual change had occurred, with male zooids occupying the small proximal thecae and female zooids in the large, often more open distal thecae.

Genetically it would be more advantageous for graptoloids to cross-fertilize because a heterozygote is fitter than a homozygote. Studies of modern hermaphrodites suggests that when sperm and eggs are produced at the same time, a genetic incompatibility inhibits self-fertilization. Such a mechanism might also have operated in graptoloids. Evidence for such cross-breeding in the fossil record is speculative. Most planktonic graptoloids were gregarious and thus there would be no physical barrier to cross-breeding. However, many species were distributed over a wide geographic area suggesting a cohesion of the genotype rather than fragmentation which might have occurred with in-breeding. *Some* of these very cosmopolitan species are relatively rare in the rock and do not typically occur in swarms on bedding planes. Furthermore such species seem to be characterised by large rhabdosomes e.g. the Wenlock cyrtograptids. If we can accept the crude correlation between colony size and longevity then we may be seeing here an expression of different reproductive strategies. The instances of genetic polymorphism, that is, when two or more forms of a palaeospecies coexisted, are thought to have been widespread, and the cause of major changes in graptoloid morphology (e.g. the change from biserial to uniserial rhabdosome). However, it is unclear whether genetic polymorphism is more likely to have occurred as a result of in- or cross-breeding.

Whatever the sexual status of the zooids throughout the life of a colony, it is conceivable that graptoloids used the two reproductive options to advantage: cross-breeding with other colonies during 'normal' conditions, and resorting to in-breeding during times of stress. To judge from the rapid evolution and proliferation of graptoloids during the Lower Palaeozoic, it was clearly a very successful method.

An unexplained phenomenon in graptolite studies is the occurrence of relatively rare *synrhabdosomes* (Fig. 110), which are groups of colonies arranged in a radial fashion. The nemata radiate from a central region; and the preservation of synrhabdosomes occurs because the nemata tangle together (there is no hard part, skeletal attachment either of one nema to another, or of the nemata to anything else other than the distal end of their own colony). Occasionally the *virgellae*, rather than the nemata, effect the radial grouping. It is assumed that synrhabdosomes were probably more common during life (i.e. they are assumed to have broken up easily during post-mortem events); and that they had something to do with either hydrodynamics/feeding or with sex.

Chapter Ten

WHAT SORT OF A LIFE DID THE BENTHONIC GRAPTOLITES HAVE?

Through the early years of research, prior to the middle of the Nineteenth Century, graptolites other than dendroids had become to some extent documented and understood, while the branched or shrubby forms (hence dendroid) were being variously described as moss (Von Brommell, 1727), plant (Hisinger, 1837), coral (Eichwald, 1842) algal (Geoppert, 1859) and bryozoan (Salter, 1866).

The first major statement about the zoological affinities of the dendroids was given by Hall in his classical memoir of American graptolites (1865). Although he had earlier (1852) suggested that dendroids were closely allied to the graptolites, it was not until 1865 that James Hall formally classified them together. He also suggested that dendroids may have had a different mode of life from the 'free-floating bodies' suggested for the rest of the graptolites and implied that the 'dendroid or tree-like' graptolites had been attached to the sea floor.

It wasn't until 1872 that the Dendroidea (or colloquially, dendroid) graptolites were recognised as a separate order (Henry Nicholson, 1872). And it was not until Gerard Holm (1890) and Carl Wiman (1895) pioneered an improvement of isolation techniques that dendroid graptolite thecal structures and their various possible functions could be elucidated. It was another fifty years or so before there was any further major development in our understanding of dendroids. That was when Roman Kozlowski (1949) was at last able to publish his major monograph on an amazing Tremadoc graptolite fauna from the Holy Cross Mountains in Southern Poland (see Ch. 8). Also extensive North American and Czechoslovakian dendroid faunas were described by Rudolph Ruedemann (1947) and Bedrich Bouček (1957) respectively. For a while little more work was done on dendroids and only recently have more papers started to appear.

Dendroid specimens are not common locally; even on a worldwide scale there are only a few prolific localities (see Ch. 4), but the fascinating nature of the fossils more than compensates for the extra effort required in finding them. Few specimens are well preserved and few species can be traced and correlated between even nearby localities. Compared with graptoloids, the dendroids are geographically restricted as species, which accords with their proposed sessile, benthonic mode of life.

Dendroids were present in the benthos from the Middle Cambrian to the Upper

Carboniferous. The time ranges of individual dendroid genera are generally also extremely long, most ranging through both the Ordovician and Silurian and some (*Dictyonema*, Fig. 26 and *Callograptus*, Figs 42 and 43) existing from the Upper Cambrian to the Upper Carboniferous. However, some of this longevity may be an artefact resulting from our lack of detailed knowledge of their morphology. As other graptolites did not appear in abundance until the Arenig and had waned considerably by late Silurian-Devonian times, it appears that dendroids were both ancestral to and outlived other members of the Class, reaching a peak of complexity during the Silurian with the *Acanthograptidae*. These last are composed of many long thecae twisting around each other, forming stipes that branch and rejoin (anastomose) in complex patterns.

The secret of the success of dendroids seem to have been their adaptability. Some species seem to have been able to tolerate local changes in environment and some were apparently able to recommence growth (Oliver Bulman, 1950) after a pause in growth or damage. The debate about evolutionary paths between dendroids and graptoloids is discussed elsewhere in the book (Ch. 12). The use of dendroids in stratigraphy is also further considered in chapters 4 and 13.

Dendroids are apparently the most complex of all graptolites. They are certainly very rewarding specimens to collect and study. The overall rhabdosome form and stipe morphology is generally clearly visible with the naked eye and much enhanced by immersion in a thin layer of alcohol. Many specimens may however show an interesting history of burial disruption, and flattened forms that were originally three-dimensional can still show a spectacular range of detail. Increased detail can be seen using a microscope, but again internal thecal structures can rarely be seen unless exceptionally well preserved material is suitably prepared (Ch. 11). It is on thecal structure and arrangement or grouping that the most reliable dendroid classification is based, but as very few specimens are well enough preserved to reveal details of their thecae, the vast majority of dendroid taxonomy is based on details of external morphology that can be observed and measured from flattened specimens. Consequently much of the classification is still at a fairly crude level of taxonomic resolution and further refinement will require better preserved material.

Dendroids consist entirely of thecae which twist and curve round each other to form the stipes of the framework of each rhabdosome. It is thought that dendroid colonies grew by adding more and more thecae to the distal periphery of the existing rhabdosome, so that almost every part is made up of thecae. The general mode of skeletal secretion was evidently the same as in the graptoloids. It is thought that each theca housed a zooid which not only constructed the fuselli of the thecae but were also to strengthen and mend their framework, perhaps even rooting it more securely to whatever it was attached, by plastering cortical bandages over the surfaces.

The thecae of dendroids are small compared with those of the graptoloids. Each rhabdosome is composed of thecae of two distinct types, the relatively large autothecae (housing ?female zooids) and the shorter, thinner and generally inconspicuous bithecae (housing ?male zooids) (Fig. 16). These are generally arranged in regular order into stipes, which branch and give the patterns so characteristic of each species. This is described more fully elsewhere (Ch. 1). The bithecae in one rather enigmatic family, the Anisograptidae, seem to suggest that it occupied an intermediate position between the dendroids and the graptoloids. The anisograptids had only a small number of primary branches and their bithecae either grew into and terminated within the autothecae, or were reduced in number and gradually, species by species, disappeared. Thecal structure and mutual arrangement of thecae varies between families. Generally dendroid thecae are relatively simple, thin, uniform tubes adapting to the shape of the stipe, unlike graptoloid thecae which are usually repetitions of the same shape and *control* the shape of the stipe. Dendroid apertures are, however, very similar throughout a colony, and are very regularly spaced, often denticulate.

Bithecal apertures are mostly simple, seen as openings on the side of the stipe (Fig. 16), occasionally facing *into* the autothecal tube. Autothecal apertures are simple openings at the end of the thecae, but the aperture can be accompanied by a swelling of the tube, a spine or an awn (protective shield). All the apertures are orientated in one direction, which in conical forms is towards the inside of the cone. Rocks containing fossil dendroid specimens always seems to break or cleave along the dorsal or anti-apertural side. It is therefore possible to have beautifully preserved and intact specimens that are too delicate to isolate, that may be equipped with ferocious apertural spines but about which we may have no information at all if the apertures point into the matrix of the rock. Only if isolated material (see Chs 2 and 11) is available, or specimens with stipes that were either torn or twisted on deposition, showing the lateral or profile view (Fig. 44), can clear details of the thecal structure then be seen. Sometimes, in the cases of (flattened) conical colonies, the thecal form can be seen at the edge of the cone where a true profile may lie in the bedding plane. In both the *Dendrograptidae* and *Anisograptidae* families, thecae are short and remain in constant contact with the stipe (adnate), reaching to or past the bud of the next generation of thecae so that stipe tips consist of only one generation of thecae at any one time. In some genera of *Acanthograptidae* and *Ptilograptidae* thecae may stand off as simple tubes to the side of the main stipe, either as individual twigs, (i.e. small groups of thecae) or as branches gradually thinning to individual twigs, down to a single theca. In these two families the thecae can be greatly elongated, reaching 1–2 cm in length, with many generations overlapping and entwining to form thick, rope-like branches of many, thin thecae.

Given the extreme variations in composition of dendroid stipes it is remarkable how constant certain dimensions are for each species. The number of thecae per unit length of stipe remains constant and, because thecal diameter is constant throughout the rhabdosome, then so is stipe width except when secondarily thickened for strength at the proximal end: similarly so are stipe spacing and angles of branching. So, in order for the cone to expand some sort of branching must have taken place, particularly for conical forms which grew by adding more thecae to the peripheral margin of the rhabdosome.

When one main stipe divided, it either bifurcates and forms two new main stipes, or a lateral second order branch forms at an angle to the continuation of the main stipe (lateral branching). For rigidly ordered genera with parallel straight stipes such as *Dictyonema*, the formation of new stipes was usually the result of *bifurcation*, often in recognisable zones at an ordered distance from the apex of the rhabdosome. Bifurcation is common to most dendroids. The formation of new stipes did not mean the demise of existing stipes, it was only the expansion of an existing stipe by the splitting of the stolon system. At a stolon node (see Ch. 1) *two* stolons were produced which went on to produce another node each; one autothecae was produced; but the bitheca was suppressed. Less common are lateral branches which are a feature of some of the less geometrically arranged genera. These lateral stipes generally caused no diminution of the existing stipe but often rapidly terminated themselves (*Thallograptus*). The nature of the stipes of each species remain constant but there is a remarkable variety of shapes of stipe, from straight unornamented linear structures (*Callograptus*, Figs 42 and 43) to twisted uneven stipes, thinning to lateral branches and twigs (*Acanthograptus*).

Connection of stipes is observed only for the two families *Dendrograptidae* and *Acanthograptidae*: stipes of the *Anisograptidae* and the *Ptilograptidae* do not reunite. Stipe connection was achieved by one of two methods; anastomosis, and the formation of dissepiments. Anastomosis is the simple process of merging of two curving stipes where their paths coincide, with the thecae briefly entwining or fused with cortex to form one stipe. The stipes may form four-sided *fenestellae* (holes in the mesh) often of a constant size and shape.

Dissepiments resemble straps reaching perpendicular to the stipes, (Figs 41 and 44) bridging the shortest possible distance between two neighbours, holding them firmly in place. Dissepiments were possibly biologically associated with bithecae and all that remains are thin tubes of cortical tissue, of constant width and spaced at fairly constant distances for each species. They should not be confused with autothecal apertural spines which are generally much thinner and were not usually involved in the process of joining two stipes. The size of dissepiment is a characteristic of each species and a whole range of sizes exists. Some species have dissepiments that are so very fine they resemble hairs e.g. *Callograptus*; others, particularly some of the robust species of *Dictyonema*, have greatly thickened dissepiments. Unlike anastomosis it is almost never possible to identify the direction of dissepiment growth; even following aberrant growth of one stipe the next dissepiment would usually have been formed at near normal spacing without evidence of having grown from one stipe to the other.

Dendroids come in many shapes and sizes. The overall adult rhabdosome size varies from less than one centimetre, up to fifty centimetres in axial length. The great range of rhabdosome shapes is based on three distinct patterns, which were themselves controlled by the thecal and branching patterns for each species. Tree or shrub-like forms result from freely branched and loosely connected stipes. Fan-shaped rhabdosomes result from slightly more regular arrangements of stipes and forms that are more sparsely branched. The most rigidly regimented species form *conical* or *funnel* shaped rhabdosomes. The extremes of this case are disc shaped, usually arising from a specimen having been squashed into the sediment perpendicular to its axis, its conical form being flattened into a disc, although a few grew in this form originally. In some cases, only one side of a conical form is preserved giving the misleading impression of a fan-like shape.

There is a great range of stipe thickness, from very robust to slender and delicate. Increases in robustness are due to a number of factors: (specific) increase in thecal size; (specific) increase in constituent numbers of thecae per stipe; and increase in secondary tissue, particularly cortical tissue around dissepiments, or round the stem and holdfasts. As has been previously discussed extrathecal epithelium (unfossilised soft tissue surrounding the stipes), may have had some bearing on the strength of the rhabdosome, or the ability of the zooids to direct currents through the mesh. If the sessile dendroid colony could only ensure a regular food supply by positioning itself in a current, then it would need to be fairly robust and flexible to withstand the buffeting that accompanied such an existence.

Associated with the strength of the whole rhabdosome is the firmness of the basal structure or holdfast. All benthic dendroids seem to have been attached to something but since the holdfasts and attachment objects are rarely preserved (Fig. 26) we know relatively little about this important aspect of dendroid growth and form. The majority of species were benthonic and a variety of different holdfast structures were evolved to safely secure the rhabdosomes to the sea floor. Holdfasts do not seem to have been specific to substrate type as one species can exhibit more than one type of holdfast in the same bed. Some forms had simple discs accompanied by either the proximal thecae fused into a basal plate, or the stipes comprised a stem with the sicula expanding at its base as the attachment disc. Some species had 'root' fibres that presumably spread out into the sediment surface in a plant-like fashion. Others wrapped around solid objects such as pebbles, or even colonised and coexisted with other animals such as brachiopods which were themselves secured to the sea floor. So it would seem that larval dendroids like the majority of living epibenthic invertebrates had little or no ability to select sites for attachment leading inevitably to some choosing insecure objects with the result that both dendroid and host were washed away.

One difference from the graptoloids is that groups of dendroids did not generally share the same basal attachment structure, although one synrhabdosome of a planktonic *Rhabdinopora* has been described. Whether this implies an ability to have actively inhibited the settlement of larvae of other dendroids on the same site is not known. One dendroid could, from one stem, seemingly support a number of separate stocks of individual conical growths (*Thallograptus inaequalis*), although this is unusual. It is tempting to equate very robust small specimens with energetic palaeoenvironments. A specimen with the combination of a robust rhabdosome and a substantial holdfast structure (or a delicate rhabdosome with a minimal basal structure), could lead us towards tentative conclusions about the turbulence or tranquility of the environment that these animals colonised. However, the relationship between robustness and energy of environment is probably not that simple. Both robust and delicate species are often found in the same beds and delicate specimens are often accompanied by very substantial holdfast structures (and vice-versa). It is likely that other factors affect this and without more information about the soft parts of dendroids we cannot be more specific. At present little is known about the particular microniches occupied by dendroids.

CHAPTER ELEVEN

HOW ARE THEY COLLECTED AND PREPARED?

(a) PRACTICAL ADVICE ON COLLECTING

Even experienced fossil collectors always feel the thrill associated with breaking open a piece of rock to reveal a handsome specimen, especially at localities where fossils may have proved to be few and far between.

So why not plan a graptolite-hunting expedition for yourself? For this purpose the minimum equipment you require consists of a geological hammer (one of about a pound to two and a half pounds weight should serve most situations), a cold chisel, a hand lens (magnification of × 10 or × 12 is preferable), toilet paper and newspaper for wrapping around your specimens to protect them from damage during transport, a measuring tape, a waterproof marker pen, a notebook and ballpoint pen to record your finds and ensure that your collections are well localised and, finally, a haversack for carrying your equipment, protective clothing and later, with any luck, your collections. Boots and wellingtons will be necessary, and a hard hat and goggles are recommended.

To start with, in order to 'get your eye in' for spotting graptolites at outcrop, it would be wise to visit one or two of the localities listed in the Appendix 2, where the abundance of specimens almost guarantees successful collection. Many professional palaeontologists have been heard to lament that they have never managed to find a graptolite! Yet they are abundant. Later you may wish to search out localities for yourself. However, it is always essential to ensure that, where necessary, you have obtained permission of access from the landowner or his representative. As fossil collectors we are all to some extent dependent on the goodwill of landowners so it is up to us to foster that goodwill. Many farmers in particular will respond helpfully and direct you to exposures if you say what sort of rock you are looking for and why.

Let us suppose you have now arrived at your selected locality. The first thing to check is that the rock face (for example, a road cutting, quarry or sea cliff) is safe. Avoid unstable parts of the section especially overhanging cliffs and remember that newly cut excavations are particularly prone to sudden rock falls. Perhaps there is a bewildering amount of rock outcrop – but there is no need to engage in wild-cat hammering. Frequently you will find that a scree of loose material has accumulated at the foot of a section, and time spent looking over such deposits can prove very rewarding, enabling you, in a relatively short time, to examine a large number of bedding plane surfaces, some of which may be nicely weathered, and thus highlighting their fossil content. Furthermore, examining scree material and noting the types of rock containing the fossils can also help you to track down the graptolite-bearing horizons in the rock face. Graptolites are often the same shade as the rock in which they are preserved, making it difficult to spot them. A wetted surface can help, so if there is a stream or pool of water nearby this can be very useful. But do not adopt this practice if it seems to be damaging the graptolites or is causing disintegration of the hand specimens as might happen in the case of friable shales. Alternatively, studying the bedding plane while tilting the specimen at various angles to the light can also assist in detecting the fossils.

If there is no scree material, then you should begin a careful study of the succession of beds, remembering what you have read in Ch. 2 about the rock types most likely to have preserved the graptolite colonies. Generally, the darker, finer-grained beds (especially black shales) are the most likely to yield graptolites. In sequences dominated by immature sandstone (greywacke), interbedded black shale bands may be extremely thin and of sparse occurrence. However, the bases of the sandstone beds themselves, or prominent bedding planes within the sandstones are worth a quick examination, as graptolites, usually current-orientated so as to lie in a common direction, are sometimes concentrated at these levels. Remember also that some graptolitic beds such as those of the Wenlock Series are characterised by a peculiar speckled appearance. This Wenlock rock type occurs in many parts of the world. Very little hammering is required at this stage, until the right rock type has been found.

Once you have found a promising bed, however, you can then concentrate on that level with hammer and chisel. The graptolites will be lying on bedding planes so that it is essential to break open the rock along such partings. Although the rocks were originally deposited as a sequence of horizontal beds, subsequent violent earth movements may have left them fractured, bent and cleaved (Fig. 39). In fact the bedding is often dipping at a high angle from the horizontal and may even be overturned, so that older beds overlie younger ones. Frequently these earth fractures, called cleavage, cut through the rocks at a high angle to the bedding. Sometimes it is difficult to distinguish the bedding planes from the cleavage fracture planes, a problem which is resolved by looking for prominent lithology changes – or graptolites! In such conditions there is a tendency for the rock to break along the cleavage planes, rather than the bedding planes – a situation which can make fossil collecting extremely difficult. One of the most consistent errors made by would-be collectors of graptolites is to fail to split the rock along the bedding plane. Rocks which are rich in graptolites may go entirely unnoticed unless the rock is correctly split.

That point enables us to explain the reasons why many workers actually fail to succeed. It is vital to find the correct rock type; it is equally vital to split the rock along the bedding plane; and it may be necessary to overcome a dominant cleavage in order to achieve this. Perhaps we can digress further for a moment to explain a trick to overcome cleavage (along which planes the rock may preferentially split). We'll describe this routine as though you were a right handed person. Firstly, wear a tough gardening glove on the left hand. Secondly, detect the bedding as described in Chapter 2. Thirdly, hold the extracted lump of rock so that the cleavage planes are horizontal, and the bedding planes dipping towards the hand holding the rock (i.e. if the rock is held in front of one, in the left hand, the dip of the bedding will be towards the left). Then take a 2½ lb hammer with a good cutting edge, and hit the upper cleavage plane with the cutting edge parallel to the strike of the bedding.

With any luck the rock then splits along the bedding, often with a tearing sound because the split is actually taking place against the grain so to speak. Be careful where you do this for both safety and palaeontological reasons. You do not want the other half of your hard won bedding plane hurtling off down a mountainside or maiming some hapless bystander; it may be the bit that has the best preserved specimens on it. In most tectonized regimes this is the only way to successfully collect graptolites with anything like the frequency needed. The rock will not split along the bedding if you lay it on the ground and beat hell out of it, even if you strike it 'on the bedding', or even if you use a chisel. Most of the Lake District, Welsh, Scottish and Irish localities fall into this category.

In untectionized terrain, of course, such trickery is unnecessary and you can lay the block on a hard surface such as a lump of dolerite or sandstone. This certainly applies in much of Shropshire and the Midland Valley of Scotland inliers, for example, as well as some of the Irish inliers.

Collections made from discrete horizons should be kept separate – remember it is easy to merge them later, but it may be impossible subsequently to separate them once they have been lumped together. It is important to take great care in wrapping your specimens. Once exposed, graptolites are generally very fragile. Considering the amount of energy that is required to break open the rocks in order to find them, it is astonishing and almost absurd how easily they can be damaged or even destroyed. A careless rub with a finger can remove much of the periderm. Consequently they have to be wrapped very carefully, first in soft tissue and then with layers of paper to make sure that they are not abraded. Give each collection a distinguishing number or letter, marking the wrapped specimens with the appropriate symbol, measure their stratigraphical distance apart, and record all these details in your notebook before leaving the locality. Ideally you should be able to relocate exactly the sites from which you extracted your collections.

Finally, please do not hammer unnecessarily and do not over-collect. Remember that others will want to come and try their luck and, for this reason, leave the outcrop in a tidy condition and ensure that your hammering and extraction of material from a rock face has not left it unstable and dangerous. If you are working on outcrops in fields with animals in them, be particularly careful not to leave sharp pieces of rock sticking out from the outcrop or lying on the ground. A particular complaint of farmers concerns animals getting cut on freshly broken rock.

(b) PREPARATION

Although isolated specimens such as were described above (see Ch. 2) give us all the more exciting information on graptolites, it is a melancholy fact that by far the larger number of specimens cannot be extracted from the rock and must be studied in the condition that nature has left them, embedded and compressed in the rock, the condition, in fact, which we mostly illustrate in this book. In practice most of the specimens you find are partly covered by rock-matrix. You can often clear this obstructive covering by careful digging with a needle – a sewing needle held in a pin-vice will do admirably – but this technique requires practice, much patience and a little luck if the results are to be satisfactory. Remember to wear goggles, needles are brittle. Furthermore, most rock contains minerals that are harder than steel, so from time to time you also need to replace the blunted needle or re-sharpen it. The technique is to take advantage of the plane of weakness where the sediment meets the surface of the graptolite itself; by digging gently into the rock it is possible to cause small flakes to lift off. Where to dig, how hard to press, above all, how to avoid spearing the graptolite itself – these are where the practice comes in. However, if the rock is not too hard to work, it is possible to expose the complete upper surface of a graptolite, though such details as spines and apertural processes are very difficult to develop out.

The cleaning and study of flattened graptolites demands the use of a good, fairly low-power microscope (most people use a binocular microscope) and, because graptolites are often black and preserved in dark rock, a strong light source. Flattened graptolites often show up best when they are wet: a film of moisture generally glosses over accidental irregularities in the rock, cuts down on specular reflection that acts as irrelevant 'noise' and at the same time enhances any colour-contrast between the graptolite and the matrix. It is surprising how much clearer graptolites can look with the help of alcohol! Water can be used for examination, but it can also damage the specimen by swelling the clay minerals in the rock and causing the periderm to flake off. In the old days clear honey was used, or glycerine; but these, though excellent for vision, are sticky and troublesome to clean away after examination. Nowadays most people use alcohol (ethanol) in the form of clear industrial meths (not the purple kind, as it has too many impurities and these leave an unpleasant goo on your best specimens when the alcohol evaporates). Ethanol does not in general affect the graptolites and it evaporates away cleanly. If you are studying a specimen at length and tire of re-moistening it every so often, a half-and-half mixture of meths and glycerine (Lovèn's reagent) gives excellent results and may later be washed off with more meths. You can make a kind of permanent wetting by varnishing the rock. This is NOT advisable because you cannot then develop the specimens, nor study them by electron microscopy. Furthermore, the varnish will probably darken with age, as has happened with specimens covered with a layer of Canada balsam fifty years ago. Nevertheless, coating with Gum Tragacanth or diluted 'Childsplay' glue may, as a last resort, be used to preserve a friable or flaking specimen. Where specimens are preserved in 3-D in the rocks, as often when pyritised, they can be studied by grinding the surface to produce serial sections of the rhabdosome and this technique was used on Swedish material by both Carl Wiman and by Gerhard Holm. A similar technique was used by Barrass on limestone specimens where he recorded each section as a 'peel'.

More complicated preparation, however, can be done on specimens isolated from the matrix as discussed above (Ch. 2). Pyritised material needs little further preparation but organically preserved specimens can be treated in a variety of ways. In his pioneer studies, Holm (1890) dissected some of his specimens with a fine needle to reveal internal structure but it was Wiman (1896) who applied the bleaching technique, initially used for fossil plants, to isolated graptolites and found that the growth lines were still often preserved. The different overlapping layers of the skeleton could be distinguished and, assuming that the growth lines represented the developing stages of the colony, the sequence of growth could be determined, confirmed by the presence of young individuals showing the relative growth stages. The bleaching was done with Schulze's solution, a mixture of concentrated nitric acid and potassium chlorate (not to be attempted without a properly vented fume-cupboard) the effective agent being oxygen produced by the reaction of these chemicals. Material from different localities reacts differently to the treatment and different species from one locality can also differ. For example, specimens from Laggan Burn near Girvan are usually quite black when isolated from the rock but amongst these *Climacograptus brevis* will clear to a pale straw colour in about 3 to 5 minutes while even very young specimens of *Pseudoclimacograptus scharenbergi* will be barely transparent dark reddish brown after half an hour. It is always necessary to watch the bleaching carefully; overbleaching will cause collapse of the specimen completely. Stop the reaction by flooding with distilled water.

Dendroids are also very slow to clear and in an attempt to improve the process we tried the palaeobotanical practice of removing the oxidised products with weak alkali. There was no effect on the graptoloids but on dendroids the result was startling. As soon as the preparation became alkaline, the dendroid began to swell up and started to dissolve. This reaction was stopped by adding further acid but one specimen which was left too long dissolved virtually completely leaving only a fragment of the stolon system behind. It would thus appear that there may be chemical differences between the collagen in graptoloids and dendroids but further research is needed as a simple structural difference might account for the differing reactions.

In the early 1920s Jaroslav Kraft discussed different bleaching methods. Initially he used 'Diaphanol' (chlorodioxid acetate) which required 1–3 days for bleaching to occur but later he suggested using concentrated hydrochloric acid with potassium chlorate which bleached his specimens in 10–20 minutes. He commented that the chlorine produced (which did the bleaching) should not be breathed in and that the metal parts of the microscope used for observing the progress of the bleaching should be protected by grease! It was probably Kraft's paper which prompted Bulman to undertake the publication in the 1930s of Gerard Holm's unpublished plates and to prepare his own material from Britain. We have also tried bleaching specimens using ordinary bleaching solution (sodium hypochlorite). The effect appears to be an overall thinning of the periderm so that specimens become very fragile and consequently difficult to mount permanently.

Since the dendroids would not bleach, Wiman studied them by making serial sections using the standard zoological techniques but his sections, at 20–25 μm, are thick by modern standards where 8–10 μm is routine. The finer structure of the stolons in dendroids can thus be followed more easily. Oliver Bulman was also influenced by the Norwegian vertebrate palaeontologist Erik Stensiö who had developed the serial grinding technique of Sollas and the preparation of models built up of layers representing each surface exposed. By making wax models of the insides of the thecal cavities, the relationships of the thecae can be worked out and viewed from any angle. The modern method of course would be to digitize the photographs or drawings of the sections and get the computer to display the reconstruction; but a wax model can be permanently displayed in the museum.

The advent of modern plastics since 1950 has made the extraction and preparation of specimens much easier, particularly where muddy sediments are involved. Before then, break up of the rocks using hydrofluoric acid had to be carried out in lead basins (very heavy) or porcelain basins coated with wax. (Beeswax made a better job but paraffin wax was much cheaper). In the great freeze of 1947 the sink outlet in the fume cupboard on the top floor of the Sedgwick Museum in Cambridge froze and Bulman had to carry the 12-inch waxed basin down three flights of stairs to decant off the HF down the outside drain before we could pick out the graptolites from the residue. Such a procedure would not be allowed nowadays on grounds of both safety and pollution!

Graptolites were usually extracted from the purer limestones by using hydrochloric acid which provided active bubbling to break up any slightly muddy layers. When acetic acid, however, was used to provide a gentler attack, some of the Swedish material yielded large numbers of conodonts and horny brachiopods appeared as well as siliceous sponge spicules. Kozlowski records the recovery of horny brachiopods and siliceous sponge spicules from the cherts that he etched with hydrofluoric acid and which yielded his exquisitely preserved graptolites.

Holm's large collection of isolated material, for which he had prepared quite a few plates, remained unpublished until it was restudied by Bulman in the 1930s. These plates are outstanding examples of illustration. For shale specimens such as the material from South America (Bulman 1931), the shale matrix was painted black and the fossil 'picked-out' in white paint before photography. For this, Holm had the assistance of G. Liljevall who also retouched Holm's photographs of isolated material and occasionally prepared crayon or wash drawings. Wiman's papers are generally illustrated by drawings except for some photographs of his microtome sections. The technical quality of much of this work is remarkably high and must have required an incredible amount of time and patience. The results can only be equalled today by the best of scanning electron microscopy.

In the introduction to the Monograph of British Graptolites (1901) Charles Lapworth discusses the best means of figuring specimens for publication and illustrates the drawing microscope which he had designed himself. Several of these Parkes-Lapworth microscopes were made, each so far as we know differing in detail, but the essential character is that the microscope is set horizontally so that the distance from the eyepiece to the drawing surface remains constant when focussing on different levels of a specimen, ensuring that the magnification of the drawing remains the same throughout. One good example is held by the Sedgwick Museum in Cambridge. Some of the original drawings for the Monograph (many of which are still in the University of Birmingham) are large, up to a metre or more long, since they were prepared at a magnification of × 5 for reduction to natural size on the plates. Lapworth stressed the need for accurate figuring – 'by mechanical means if possible' – so that even the imperfections of the specimens should be shown, and also the uniform scale 'for the purpose of easy comparison'. The only preparation of specimens figured in the Monograph was the mechanical removal of overlying matrix since most of them were preserved in typical black graptolitic shales. The majority were from existing collections although work in Central Wales in the early 1900s while the monograph was being published resulted in some fresh material. Photography of such material was difficult with the methods then available and publication of photographs at natural size would have been impossible. By having enlarged drawings, with some details of texture applied as a wash, photographed on to collotype plates, the resulting natural size figures could be studied with a hand lens when comparing them with actual specimens and real comparisons made. The text figures in the Monograph were similarly prepared at × 10 and reduced to × 5 for publication. The Parkes-Lapworth microscope was used by Ruedemann for some of his work on North American graptolites and both Sun and Hsü used the same technique for their 1930s papers on Chinese graptolites.

Techniques of photography and film quality have improved during the last 50 years and it is now generally possible to produce usable photographic prints. However, there are apparently still problems with the publication of photographs and most papers from the 1940s onwards have also used simple line drawings to illustrate the shale specimens with photographs to supplement them. There is often a considerable difference between the photograph of a specimen taken dry and one taken under liquid as well as the problem of directional lighting (Figs 10 and 11). Few specimens are absolutely flat but high-angle lighting is usually needed to reflect off the specimen. This can be achieved by ring lighting but the result can look unnaturally flat. A separate beam to highlight the specimen is then required and balancing the lights can take a lot of time.

The photography of isolated specimens presents different problems. The generally small size (often only 2 or 3 mm long but up to 1 mm thick) of the specimens involves using a microscope objective lens with a very small depth of focus and the reddish brown colour meant that exposures were very long before the development of panchromatic film. Kraft produced some colour photographs and also considered the use of infra-red film which enabled clearer pictures of the growth lines to be obtained. Because of the difficulties it is not surprising that Kozlowski (1949) and Bulman (1947) still chose to illustrate their studies by wash drawings. Most of Kozlowski's specimens were unbleached and a carefully shaded drawing shows what can be seen rather better than ordinary photographs. Bulman's Laggan Burn material includes many bleached specimens and his wash drawings give the impression of photographs. It is possible, in a drawing, to highlight points in a way which is impossible in a straight photograph. For the same reason much medical illustration is still done by artists rather than cameras. Some of Bulman's earlier line drawings were prepared by inking in the lines on a photographic print and then bleaching out the image to leave a drawing suitable for a line-block which was the standard way of printing text-figures. The development of offset-litho printing has removed this limitation and made the printing of photographs much easier. The advent of the scanning electron microscope (SEM) has also changed the type of illustration available for isolated material as the pictures have a great depth of focus. However, they only show the surface details of the specimen and not the detail seen in a bleached specimen through different thecal walls. The magnifications in the SEM were also generally too high so that a succession of prints had to be assembled into a montage with possibilities of distortion. It is also not always easy to relate the small part of the rhabdosome seen on the SEM screen to the whole specimen.

For her studies on the nature of the retiolitid skeleton, Nancy Kirk has produced large scale models of the meshwork so that it can be studied more critically. Greatly enlarged models of *Climacograptus inuiti* were made by Ian Cox in 1934 to show the development of the rhabdosome with the successive thecae in different colours, and Nancy Kirk's models of retiolitids have also been coloured for ease of interpretation.

FURTHER INFORMATION ON EXTRACTION AND ISOLATION

Well-preserved three-dimensional graptolite specimens can be found in fine-grained limestones, shaley calcareous mudstones, and cherts; such specimens can often be completely freed from the rock for full study. It must be emphasised that, although suitable specimens may be abundant at a particular locality, such occurrences are in general very rare. Nor would many of such graptolite specimens stand up to the harsh treatment about to be described.

Small pieces of rock are treated with dilute acid (hydrochloric or acetic) to first remove the carbonates. If the rock is more or less a pure limestone, this may be all the acid treatment necessary to completely free the graptolites (Fig. 133). Often, however, further acid treatment is necessary. After several changes of acid over a period of three weeks, ensuring complete dissolution of the carbonate, the remains are then treated with 60 per cent hydrofluoric acid (in polythene dishes, and safe at the back of the fume cupboard; it cannot be stressed too often that this acid is odourless, colourless and highly dangerous!) to remove the silica content. What remains are the graptolites themselves and some muddy sediment. The acid is thoroughly washed away with several changes of distilled water, when the graptolites can be picked out, using a wide-mouthed pipette, into dishes and stored in glycerine: in this condition they can be stored for a hundred years and more. Alternatively, they can be further treated for study. Hydrofluoric acid can result in a quick death, so great caution must be shown and protective clothing worn.

Usually the graptolites are black and opaque, a result of the carbonisation of the rhabdosome – for light microscope study they need to be made translucent (Fig. 131). This is done using Schulze's Solution (concentrated nitric acid and potassium chlorate). When they have become translucent, they are dried in alcohol and cleared – using xylol – and can then be mounted on glass slides in Canada Balsam or retained in glycerine. Use of a light microscope will then reveal the structure of individual thecae and the sicula, often clearly showing the fusellar rings. Some remarkable wash drawings by Kozlowski published in 1949 even show what we now know to be cortical bandages (e.g. Fig. 19).

Many of these isolated specimens are not robust enough to support their own weight in air, but if they are strong enough to do so then they can be examined with a scanning electron microscope (SEM). For this, the surface of a SEM stub is coated with a suitable adhesive, such as Photofix or double sided Sellotape, and the washed graptolite is transferred in a drop of alcohol to the stub. The alcohol evaporates, leaving the graptolites stuck to the stub. It is at this stage that one finds out just how robust the specimens are! It is quite unnerving to find apparently well preserved three dimensional specimens collapsing down on themselves as the alcohol evaporates. For example, *Monograptus priodon* from limestone nodules of Cornwallis Island, Arctic Canada, are extremely robust and can stand any amount of this treatment; contrast this with the delicate specimens of *Glyptograptus* spp. from the Ural Mountains, USSR, which flatten out completely when mounted on stubs. Once mounted, the stub is then coated with a very thin layer of gold. SEM study can then reveal the most incredible detail of fabric structure, not only of the fusellae and the cortical bandages, revealing individual fibrils within the cortical bandages (Fig. 132), but also of sheet fabrics forming granular layers over and within the other fabrics.

Other well-preserved graptolites can be embedded in epoxy resin and cut into ultra-thin sections (ideally about 600Å/60 mm thick) using an ultramicrotome with a diamond knife. Lori Dumican and Barrie Rickards (1985) give details of the preparation of the graptolites and the actual method used. The ultra-thin sections float off from the cutting edge of the knife on to the surface of a water bath, from where they are picked up on a copper grid after being flattened with a whiff of chloroform. When the grid has dried it is then ready for examination under transmission electron microscope (TEM). This will then reveal details of the internal structure of the different fabrics which compose the periderm (Fig. 132). It shows fusellar fabric to be composed of fibrils in a fairly open, spongy meshwork, bounded by a thin, dense layer of sheet fabric. Cortical fabric consists of closely packed parallel fibrils, with individual cortical bandages having the fibrils running along the length of the bandage. This gives the effect in TEM of layers of fibrils running in different directions, with thin layers of sheet fabric between the layers of cortical fabric.

Chapter Twelve

HOW ARE THEY CLASSIFIED?

The classification of organisms is often misconstrued as a static and boring subject. It is not. Classification may be complex and difficult but it is also dynamic. Without it we could hardly begin to study what is perhaps the most important topic in biology, if not the whole of science, and that is evolution. The identification of graptolites down to species level is quite a skilled matter; and there is more to identifying fossils than just giving them names. The process of identification also inevitably involves an understanding of the details of the ways the fossils are constructed. This understanding is always changing, and as a result there is nothing invariable about identification. Names can change as we get to know more.

This fact is reflected in the history of the identification of graptolites. As we have seen (Chs 1 and 10), there was at first little notion of what graptolites were. Despite lack of knowledge about their fine structure or growth they still made distinctive patterns on the rocks. Consequently they could be used stratigraphically to distinguish various horizons. A quite sophisticated stratigraphy was developed in the mid-nineteenth century before most of the modern ideas about graptolite evolution had been developed. James Hall, (see p. 142) the great North American palaeontologist, named dozens of species in the 1850s and 1860s. Most of these he placed within a single genus, *Graptolithus*. In the ensuing decades Charles Lapworth, (see p. 143) Henry Nicholson, Sven Törnquist and others laid the foundations of our current classification by 'splitting off' and naming many more genera, and the old catch-all genus *Graptolithus* disappeared forever. The foundations of the modern graptolite families date from Charles Lapworth's (1873) review.

Nonetheless it remained true that the genera proposed in these early days were mostly based on what one might term the gross aspect or silhouette of the colony, usually as seen flattened. The identification of the genus entailed the kind of observations that can often be made in the field – with the help of a lens. The features involved would be: the number of stipes and their arrangement: whether or not the stipes were scandent (i.e. back to back, e.g. Fig. 99); and to some extent the style of the thecae (Fig. 129). Species might then be distinguished back in the laboratory by such features as the thecal 'density' (i.e. number in an inch) in the 'polypary' (most authorities thought graptolites were hydroids then), and such subtleties as the angle between stipes and their overall size. Many of these characters are still considered important or useful today.

The beauty of this early classification system was its practical utility. It did not take long to master the names of the thirty or so graptolite genera. Furthermore the names were often easy to assign because many of them were obviously related to the form of the fossil. *Tetragraptus* had four stipes; *Didymograptus* two stipes; *Monograptus* but a single stipe. What could be easier?

As this naming system was being elaborated it was also becoming clear that the stratigraphic distribution of graptolites matched the names wonderfully well. Armed with his vocabulary of names, the student could look at a few slabs covered with graptolites and not only name most of the genera, but give a reasonable account of the age as well. This remains true even today. The many-stiped graptolites were dominant in the earlier (Tremadoc, Ordovician) faunas. There was a phase somewhat later when few-stiped *Tetragraptus* and *Didymograptus* predominated; then came faunas dominated by biserial graptolites which would have been placed in the genus *Diplograptus*; and so on through the Ordovician and Silurian, culminating in 'the age of monograptids'. No wonder that the graptolites came to be regarded as the ideal stratigraphic tools: they were evidently comparatively simple to identify, even in the field, stratigraphically limited, and found all over the world in the same order. The sequence of graptolite faunas came to be seen as closely reflecting their evolution; genera with their handy characteristics rose to prominence and declined, to be replaced in turn by others. In this way stratigraphy and classification became intertwined. The evolutionary history of the graptolites could be read from the paper shales like the plot of an historical novel.

To some extent this attitude is still in currency today. The fact is that graptolites *are* among the most widespread of fossils, and *do* occur in consistently similar order, often nearly worldwide. There are large areas of Lower Palaeozoic rock, such as the Southern Uplands of Scotland, whose complex structure in no way could be understood but for the relative dating and sequencing provided by graptolites. The classification that resulted might be described as 'horizontal' with respect to time, in that, rather like the chapters of a book, the classification units tended to be rather neatly chopped off into chunks bounded by particular time planes. These time planes corresponded with the old divisions between successive faunas characterised by particular genera recognised by the early students of the graptolites.

But this simple view of identification from the gross aspect of the graptolite colony hardly applies today. At first sight it seems a pity, if only because the old system had such simplicity and elegance. The business of identification has become a more specialised matter, and this often seems good only to the specialist! However, the change was inevitable as more was discovered about the detailed structure of graptolites: knowledge changes classifications. The investigation of graptolites isolated from their rock matrix, pioneered by Holm and developed by Kozlowski and Bulman, really provided the first insight that all was not well with the old graptolite genera. As more and more graptolites have been investigated in this fashion, so it has been found that the old ways of recognising genera by stipe number, say, or even thecal shape, are not really appropriate. The proximal part of the graptolite can as a rule only be satisfactorily seen in isolated specimens, and it is this proximal part which reveals some of the most fundamental details about the evolution of graptolites. It was here that branching was initiated, and here that the various patterns and styles of budding determined the overall form of the colony. The first few thecae are, naturally, 'built in' to the rest of the colony, so that it is possible to identify species with similar proximal end structures. This work is enormously laborious compared with simply squinting down a lens in the field, or counting the number of stipes. But what was revealed was important. Species with similar 'gross aspect' – which would have been placed in the same genus – often showed fundamentally different structures at the proximal end. Worse still (or better, depending on your point of view), the same kinds of proximal end structures could be found in other species which would have been placed in different genera on gross aspect. What we are seeing, of course, is parallel evolution.

It did seem probable that similarity in proximal end structure was more fundamental to classification than other aspects of general colony similarity. After all it was a relatively simple matter for a *Tetragraptus* to 'lose' one or more branches to become a *Didymograptus* because this requires only one or two decisions in the growth of a colony, whereas to completely restructure a proximal end requires numerous decisions in growth and form. Besides, it is a general biological rule that growth decisions taken early on in the growth of an organism are more refractory to change.

What this meant was that many of the old genera were probably not natural units. A *Didymograptus* species with a distinctive proximal end may be more closely related to a *Tetragraptus* species with the same proximal end, than to some other (erstwhile) *Didymograptus* with a different proximal structure. *Didymograptus, Tetragraptus* and many of the old, familiar names were merely 'form genera', serving as useful packages to wrap up species, but lacking phylogenetic meaning. The days of the instant field identification are largely over. Even such familiar genera as *Phyllograptus* – the 'leaf graptolite' – are likely to have been derived from more than one phylogenetic source. The old genera have now to be employed in a more limited sense around the species on which they were first based, although they can still be used *sensu lato* for flattened specimens. This means that genera are 'split' into finer units which more closely reflect evolution: ideally, a genus will include an ancestral species and its descendants.

What this means is that the former stratigraphic classification, which was described as 'horizontal' above, is being replaced by a 'vertical' one, in which individual phylogenetic lines are being traced through time, in part cutting across the former chapter endings that divided graptolite history into easy chunks. The old genera were for the most part representing a *grade* of organisation (e.g. four stipes), whereas what classification requires is a *clade* (a single branch of the evolutionary tree). The old form genera acknowledge that a similar grade of organisation was reached by more than one clade at more or less the same time in graptolite evolution: parallel evolution. This is one of the most compelling facts about the evolution of graptolites. This interaction between grade and clade is only really being worked out at the moment. Graptolite classification is therefore in a state of transition, with the old form genera persisting alongside new concepts. With few exceptions, authors now are trying to make their classifications phylogenetic ones. One might, perhaps, summarise the characters used to classify and identify graptolites in the following order of importance (note that the old criteria are still there, but with reduced emphasis).

1. Structure and development type of the proximal end.
2. General habit (scandent, uniserial etc.).
3. Thecal type (climacograptid, isolate etc.).
4. Branching type and arrangement (if branched).
5. Thecal spacing, pattern of stipe expansion, stipe width, and other quantifiable aspects of single stipes.
6. Overall colony size.

Read the other way round, this list would be one reflecting the ease with which characters can be changed. Overall colony size could have been changed, even ecologically, by differences in nutrition, or some small genetic perturbation, whereas proximal structure would tend to be conserved. Species tend to be defined by differences near the bottom of the list (so you can find fat or thin, large or small species of the same genus), families or higher categories by characters near the top of the list.

It will not always be possible to see all these characters on fossil material – particularly a poor specimen preserved in shale. Quite often it is necessary to make an educated guess about the fundamental proximal structure of such species. None of these advances in classification have affected the original observation that a general age can be obtained by looking at the aspect of a fauna 'on the rock'; the *grades* of organisation through time have remained broadly the same.

PHYLOGENY (see also Appendix 4)

The Class Graptolithina includes a number of encrusting groups usually classified as Orders (Crustoidea, Tuboidea, and Camaroidea) (Fig. 1). These curious organisms are not widely known, indeed many of the species are recorded only from the Ordovician of the Baltic – whence they were spread over Europe, within boulders, during the Pleistocene glaciation. Like the dendroid graptolites, these groups had two sorts of thecae with apertures: autothecae and bithecae. It is assumed, perhaps wrongly, that the presence of such thecae is an advanced characteristic of the Graptolithina, as is the presence of fusellar tissue. The relationship of the encrusting orders to the Rhabdopleurida (Fig. 1) on the one hand, and the Dendroidea on the other is not resolved. Presumably, by analogy with *Rhabdopleura*, the encrusting habit was primitive for these Hemichordata.

The Dendroidea plus Graptoloidea together constitute a natural group, and are of the most immediate concern to this book, as they are by far the commonest hemichordate fossils. The Order Dendroidea has a shrubby habit, and differs from encrusters by having a stem. Their classification and identification is unsatisfactory, because it is still largely based on the overall shape of the colonies (Ch. 10). A detailed knowledge of a few isolated dendroids is sufficient to tell us that they can be very complicated animals indeed in their colony construction, but few specimens are well enough preserved to show such details. The stratigraphic ranges of genera, some ranging from the Cambrian to the Carboniferous, indicate that we have not really begun to understand how to divide the group phylogenetically – or else that they are the most conservative of all Palaeozoic animal groups, with the possible exception of the brachiopod *Lingula*. The former explanation seems more likely.

The Order Graptoloidea is characterised as including those Graptolithina retaining a nema. They are also those which had planktonic habits, and are the 'graptolites' useful in correlating rocks in common geological parlance. The family Anisograptidae includes the most primitive Graptoloidea, those retaining bithecae. They have long been recognised as temporally and morphologically transitional with dendroids, with which they have usually been classified in the past. The Graptoloidea were then defined by the lack of bithecae. However, Richard Fortey and Roger Cooper (1986) showed that this definition of Graptoloidea was unsatisfactory, not least because there was good evidence that bithecae were lost on more than one evolutionary lineage. If defined thus the Graptoloidea would have been polyphyletic, which is not acceptable (this is another case where a former 'horizontal' classification has given way to a 'vertical' one – the Graptoloidea used to 'start' at the Arenig base). Most authorities seem to agree that the Graptoloidea acquired planktonic habits only once, and that they are descended from a single common ancestor. In most rock sections the conical *Rhabdinopora* (Fig. 41) (in part *Dictyonema*, Fig. 26, of older texts) is the first planktonic genus encountered at or near the base of the Ordovician (Tremadoc Series), and it has been assumed that *R. flabelliformis* was the *fons et origo* of all the Graptoloidea. There are now some other possibilities, and the conical form has been regarded as something of a side-branch in graptoloid evolution. Whatever the case, the earliest graptoloids had very rapid dichotomies at the proximal end to produce what is described as 'four primary stipes' – giving a cross-shaped appearance in the flattened state. This was very quickly reduced to three primary stipes and then to two, which then remained the case for virtually all subsequent graptoloids, until the proximal dichotomy was lost in the uniserial Monograptacea.

The loss of bithecae occurred in three, or maybe more, groups of graptoloids at or close to the base of the Arenig Series, thereby marking the arbitrary end of the anisograptid grade. The ensuing Arenig Series included the first representatives of the three major groups of younger graptoloids: Dichograptacea, Glossograptacea and Virgellina. All these groups produced genera with scandent habit; the last named alone survived the Ordovician, and, having nearly become extinct at this time of crisis, survived to blossom during the Silurian and Devonian until the extinction of the graptoloids.

Dichograptacea include many of the genera which have been taken as being typical of the Arenig, with dichotomously or laterally branching rhabdosomes. Several evolutionary lines involving reduction in stipe number lead to rhabdosomes with *Didymograptus* (two-stiped) grade of organisation. *Didymograptus* itself includes the familiar Llanvirn 'tuning fork' graptoloids. The scandent tendency is illustrated by the four-stiped *Pseudophyllograptus* – essentially a *Tetragraptus* with the four stipes lying back to back. The stipe reduction series reaches a maximum with *Azygograptus* (Fig. 119).

Glossograptacea are two-stiped graptoloids with large thecae, and a characteristically symmetrical arrangement of the proximal end. Many became scandent. The horseshoe-shaped isograptids typify the later Arenig to Llanvirn in the Ordovician, while the glossograptids are found in Llanvirn and younger Ordovician rocks. These latter are superficially like the diplograptids, but their structure shows that they are not closely related. The most curious glossograptacean is probably *Corynoides*, a form which became mature with only sicula and two completed thecae – but these are gigantic. This is probably a case of 'arrested development' (progenesis).

The Virgellina include by far the largest morphological and stratigraphical range of graptoloids. They are characterised by a stiff rod (virgella) extending from the sicular aperture. The first members of the group appear in the Arenig. Biserial graptolites with two series of thecae back-to-back would originally have been accommodated within the form genus *Diplograptus* by workers in the mid nineteenth century, but are now placed in many different genera according to the details of their development. They range upwards into the Silurian (Ludlow). Mid- to late Ordovician rocks are characterised particularly by dicranograptids and nemagraptids in which all or part of the stipes are again uniserial. At the end or the Ordovician nearly all graptoloids became extinct, except for a few biserials, and these (in the *C. extraordinarius* Zone, Fig. 98) must have included the ancestors of all later virgellines, including monograptids. The first monograptid *Atavograptus ceryx*, is much like a biserial in which one series of thecae has been repressed. The monograptid evolution in the Silurian is famous for its rapidity and stratigraphic utility. The old form genus *Monograptus* is now greatly subdivided into smaller units, based on, for example, the thecal form, which in monograptids can become fantastically lobed, hooked, twisted or extended. Some biserial virgellines evolved the net-like reduction in periderm to produce the delicate colonies of retiolitids and allied forms. Late virgellines produced intriguing homoeomorphs of early (Ordovician) branching genera; but the branches are not produced by regular dichotomies as in primitive graptoloids, in which one theca gives rise to two daughters, but by the generation of *cladia* – branches produced from the apertures of thecae or siculae. Clearly even at the end of their history there was a living to be made as a branched graptoloid. Monograptids developed the spiral design in a way that had not been done by many earlier graptoloids (*Cyrtograptus*; Fig. 58). The latest monograptids include some fairly simple forms, however, and it is not possible to invoke some kind of 'over specialisation' to explain their disappearance in the Devonian.

EVOLUTION (see also Appendix 4)

The gross evolution of graptolites was illustrated in Fig. 1. It seems very likely that in the earliest Middle Cambrian an (extant) group of hemichordates (namely the pterobranchs) separated from the Graptolites proper, for both began in the Middle Cambrian. At least, on present evidence they did not occur earlier. At present there is a dearth of evolutionary information available that might allow us to link a number of the graptolite orders, a situation which largely reflects a poor stratigraphic record of all the orders except the Dendroidea. In terms of pure morphology there is a clear similarity of some orders of graptolite with the rhabdopleuran hemichordates, to the extent that recent research has necessitated the almost wholesale transfer of forms from one group to another (Pietr Mierzejewski, 1986).

There is some scope here for a cladistic analysis (which has not yet been carried out) because the stratigraphic information is no real help in this area. Thus the actual origin of graptolites is in doubt, a situation summarized by Barrie Rickards (1979).

Related to this rather difficult situation, ripe as it is for further research, is the fact that with these very early forms a decision often needs to be made as to whether or not a fossil is a graptolite at all or a member of some other hemichordate group or even a hydroid. The criteria outlined in Chapter 1 are no help because the rhabdosome and stipe form is atypical – intuitively we note that it is 'primitive'. What this really means is that structural details, such as thecae and fuselli, are often not detectable, partly because they are not well preserved in such ancient rocks, and partly because they are not present in quite the same form. *Mastigograptus*, for example, only rarely shows its autothecae, but does show a black, 'structureless' stem from which autothecae and bithecae may or may not develop.

Thus the early evolution of graptolites is really tied up with the difficulty of deciding whether or not the fossils collected are actually graptolites. In order to determine this it is necessary to identify fuselli and, preferably, thecae. The nature of fuselli should readily distinguish graptolite from other hemichordate forms, and the presence of fuselli, however moderately preserved, does at least confirm hemichordate affinities.

There are a number of well preserved dendroid genera present in the Middle Cambrian (e.g. *Dictyonema*; Fig. 26) and in these forms we can be fairly sure from the branch and thecal geometry that we are dealing with graptolites *sensu stricto* even in the absence of fusellar growth lines. In summary one can note that there is considerable hemichordate diversity in the Middle Cambrian, with most of the major groups established, but that evolutionary relationships are still necessarily unclear.

In the Upper Cambrian a not dissimilar situation obtains except that the 'normal' graptolite genera – *Dictyonema, Callograptus, Dendrograptus, Desmograptus* etc. are more strongly represented, along with, a still considerable variety of the more obscure groups (see Barrie Rickards *et al. in press*) such as *Archaeolafoea* and *Aellograptus*. There has been a recent claim that the first planktonic graptolites were Upper Cambrian but our re-examination of all the type and figured specimens suggests that there is no evidence for this assertion.

In fact, the earliest unequivocal planktonic graptolites were earliest Ordovician: the family Anisograptidae is intermediate between dendroids and graptoloids, and cogent arguments have recently been put forward by Richard Fortey and Roger Cooper (1986) to place this family firmly in the order Graptoloidea. In any event they were the first planktonic graptolites on earth, the first with cosmopolitan distribution at specific level as a general factor, and the first group where we know a great deal about the internal development from chemically isolated collections.

It is at this stage, in the Ordovician, where general appearance of the colony is sufficiently distinct to enable the label 'graptolite' to be used with confidence. The spectacular geometry of *Clonograptus, Loganograptus, Dichograptus, Goniograptus* and many others (Figs 2, 63, 64, 81 and 112) brooks no argument in this respect. So identification of graptolites as graptolites becomes easier as one deals with higher stratigraphic levels. It is perhaps ironic that the details of fuselli are also easier to ascertain at these higher levels in the stratigraphic column.

With the planktonic mode came a great diversity of rhabdosomal geometry, and a consequent recognition of many species. This is shown diagramatically in Fig. 1 as a series of evolutionary bursts in the Graptoloidea. Within these 'balloons' many of the usual evolutionary patterns are identifiable such as mosaic evolution, parallel evolution, convergence, polyphyletic origin of earlier described genera, and so on. Indeed, the stratigraphic record is, particularly in the Silurian, so good that a zonal resolution of less than 1 Ma is possible, and analyses of the evolutionary patterns contribute greatly to general theses on evolutionary mechanisms (e.g. echoic and heraldic evolution in Barrie Rickards, 1988).

The major evolutionary stages in the Ordovician encompass loss of bithecae in several lineages, resulting in a polyphyletic Dichograptidae; a reduction in number of stipes (except for, secondarily, the addition of lateral branches); an increased rhabdosomal diameter; the development of planktonic morphological features (nemata, vanes, webs, spinosity, peridermal reduction); a change from pendency, to horizontality, to reclination and scandency and the origin of the compact biserial graptolite (e.g. Fig. 99). Many of these lineages, or at least the gross representation of them are illustrated in Appendix 4 with cross reference to the relevant illustrations.

In the Silurian and Devonian periods the same tendencies continued but the full range of morphological diversity became to some extent restricted, the faunas being dominated by uniserial and biserially scandent forms. However, in the development of thecal and sicular cladia there was a mimicking of earlier (Ordovician) rhabdosomal types (e.g. *Nemagraptus* v. *Sinodiversograptus*; *Pleurograptus* v. *Abiesgraptus*). By the onset of late Silurian times biserial forms had disappeared, and most uniserial scandent forms (*Monograptus s.l.*) were relatively short. A few genera, such as *Linograptus* and *Abiesgraptus* were multistiped, with thecal and sicular cladia. Many of these Silurian and Devonian lineages have been studied at specific level (as in Barrie Rickards and colleagues, 1976) and the actual demise of the planktonic graptolites has been summarised by Tania Koren' and Barrie Rickards (1979). In contrast to the Cambrian graptolites in general, and later dendroids, the Ordovician to Devonian graptolites (the plankton) are well monographed and often readily identifiable at specific level, and it follows from this that details of, for example, thecal morphology are well known. Nevertheless it should be emphasized that a conservative estimate of the number of species about which little is known other than a general silhouette, would not be less than 50 per cent: so there's a great deal of research to do. Add to that the spectacular discoveries taking place each year and there is clearly no end to the research needed nor, indeed, to the increasing potential of graptolites in evolutionary and biostratigraphic work.

Finally, your attention should be drawn to Appendices 5 to 7 which explain how to use the museum comparative collections, how to find the literature, and how to find people who will be able to help with the nitty gritty of putting an actual name on the fossil you have collected.

Chapter Thirteen

WHAT USE ARE THEY, ANYWAY?

Interesting as they might be as organisms in their own right, what use are graptolites and why should palaeontologists spend time studying these long-extinct and somewhat obscure fossils?

It has long been recognised that fossils provide a means of understanding the order in which beds of rock were deposited. Where layers of sedimentary rocks are clearly seen and are relatively undisturbed, it is often possible to observe that the fossils in the lower, older rocks differ from those in the higher, younger rocks. It may also be possible to show that a whole range of plants and animals appeared and disappeared as time passed and as you move from oldest to youngest rocks. Such a sequence of fossils forms the basis of *biostratigraphy*, the ordering of rock layers by means of fossils.

Having established a biostratigraphic sequence, isolated outcrops of rocks can be related to that scheme by means of the fossils found there. In other sequences, it may also be possible to show that some layers of rock are missing. This relating of rocks from one place to another is known as *correlation*. We see, therefore, that fossils provide a key to correlation. Even if the type of rock differs from one place to another, for instance a shale in one place and a sandstone elsewhere, if the fossils are the same we may consider as a first approximation that the rocks are of more or less the same age.

But why all this fuss about correlation? Aren't there more important things to do with rocks, such as working out what they are made of, and what they might contain that is of use, or of harm, to people? The answer is that these things are extremely difficult to do without knowing how any particular, local sequence of sedimentary rocks fits into the overall geological framework of a region. This 'fitting-in' generally cannot be done by simply following physical layers of rock until they join up. For a start, there is the problem of the superficial layers of vegetation, soil, boulder clay, discarded brass bedsteads, dead sheep and the like, which often obscure the 'bedrock' as it is called. Because of this, not more than 10 per cent of rock layers can generally be seen and more often than not they are crumpled into folds, and chopped up by faults – as can be well seen in the Southern Uplands of Scotland, or the Lake District. And, even without such complications, layers of rock do not stay the same when they are followed in any particular direction.

There is, of course, no reason why they should. They just represent the nature of the sea floor at the time of sedimentation and as the sea floor is not particularly uniform at present, so the rocks change to reflect past changes in the geography of the sea floor. It is only by understanding the way these ancient geographies fitted together in time as well as in space that we can try to work out where the useful – and the harmful – materials within the rocks are likely to be found. This 'matching-of-the-age-of' rock sequences is generally done by examining the fossils that the rocks contain. This only gives a 'relative age' but in practice that is all that is necessary. Direct or 'absolute' dating of rocks, using the radioactive decay of certain isotopes of uranium, potassium and other elements to give a figure in millions of years, is only possible for certain types of rock (most of which are not sedimentary). Not only that, it is time-consuming, difficult, expensive and sometimes fraught with error. For most practical purposes of correlation, *fossils are the thing.*

However, not all fossil groups are useful in correlation. Obviously any fossil that is so rare that it is unlikely to be found except by luck will be of little practical use. Dinosaurs, for instance, have their admirers, but they are not often used for correlation. Similarly, a very common fossil species which existed for a relatively long time will not provide the precision needed. The brachiopod genus *Lingula*, which has persisted for some 500 million years with very little change, makes a strong claim to be the world's most useless fossil in this regard.

To be of use in correlation, a fossil group needs to fulfil a number of criteria. First, its species need to be common enough to be found fairly easily. Next, they need to have a widespread distribution. And last, they need to have evolved rapidly. Many fossil species or groups prove useful in correlation, but most of the macrofossils which we find are of shelly organisms which tended to be restricted in their distribution, having been dependent in life on a particular type of sea-floor environment and often limited by sea-floor depth or water temperature.

Graptolites on the other hand are, in theory and often in practice, close to the ideal fossil for use in correlation. For one thing, they appear to have evolved quickly. Now, an awful lot of time can be spent debating the extent to which biological species and biological evolution can be recognised in the fossil record, given that we only have the hard parts to work on. Undoubtedly with many groups of organisms, much of the evolutionary story simply hasn't been preserved. Changes in the intricate colonial skeleton of the graptolites, however, probably reflected the group's evolution particularly well. Further, rates of evolution, especially in the Silurian, appear to have been relatively rapid. It seems that different types of organisms do not always evolve at the same rate. In the last few million years, for instance, some groups (e.g. the molluscs and beetles) have undergone little or no perceptible evolutionary change, give or take the odd extinction. In the same time, other groups (the mammals in particular) have evolved, undergoing several speciation events within some lineages. The reason for this is not very well understood; some seemingly important factors, such as length of reproductive cycle and sensitivity to environmental stress, have apparently played little part in this phenomenon. During the Silurian, the graptolites, for whatever reason, seem to have been the evolutionary pace-setters of the animal kingdom.

Because they mostly lived a planktonic or pelagic life within the sea waters, rather than on the sea bed, their distribution is much more widespread than most shelly fossils. For this reason too, graptolites occur in rocks where most other fossils are absent: rocks deposited in deep seas or far from shore where the sea bed was too inhospitable to support life. These kinds of 'deep-sea' rocks are very widespread, forming, for instance, much of central and north Wales, the Lake District, the Southern Uplands of Scotland and several parts of Ireland (see also Ch. 4). The graptolites in the waters above them would eventually perish and sink to the sea-floor where they remained in the slowly accumulating sediment undisturbed by scavengers. Some graptolites would also be preserved in sediments much closer to the shore in shallower deposits amid shells and other fossils. A similar mixing occurs where shells have been swept by currents into deeper water environments and such occurrences provide a link, therefore, between the graptolite-based biostratigraphic sequences of the off-shore environments and the shell-based biostratigraphies of the near-shore shelf sea deposits.

The free-floating mode of life of most graptolites also led to a widespread (or *cosmopolitan*) geographical distribution. Clearly a number of factors which varied with time will have influenced this geographical distribution: ocean currents, land barriers and the degree of tolerance which different graptolites had to changes in ocean temperature, all played their part. Nevertheless, in the Silurian especially, graptolites tended towards cosmopolitanism or world wide distribution.

Graptolites also gain an advantage in correlation as a result of their planktonic lifestyle. All other types of colonial animals that are commonly preserved in the fossil record lived attached to the sea floor, where they could show an alarming ability to change their shape to fit in with local environmental conditions. A single species of coral on a reef, for instance, might be fragile and branch-like in form, where it is sheltered, but rounded and compact on the exposed, wave-swept part of the reef. This can lead to some perplexity when trying to classify the animals into species using the skeletal parts – the only bits that are normally fossilised. Not so with the graptolites: out in the ocean waters, where environmental stresses are much more uniformly distributed, individuals within any one species of graptolite developed skeletons of strikingly similar shape, with very little variation around the 'average form'. This makes the task of identification considerably easier.

So, graptolites provide a widespread, relatively common tool in the task of correlation, but how precise can they be in giving an age for their host rocks? There is more known about the evolution of the graptolites in some parts of their history than others, but overall, enough is known about the rise and demise of different species and of the assemblages of species that may be found together to allow fairly precise estimates of relative age. In the Silurian, for instance, biostratigraphers recognise some forty graptolite zones. That is, forty zones each of which is characterised by a particular species or, more usually, an assemblage of species. The Silurian itself is estimated to have lasted between 20 million and 30 million years, so on average each zone had a duration of less than one million years. Individual species persisted from less than half a million years to five or more million years; a typical 'species-span' might be about 2 million years. It is often possible to relate graptolites or assemblages to say the lower, middle or upper part of a zone. Thus, precision to within some hundreds of thousands of years is possible. The Silurian lasted from 428 million years to 408 million years before the present. The potential precision in dating Silurian rocks relative to the present by means of graptolites may therefore be in the ratio of about 400,000:400,000,000 or to *within* 0.1 per cent error! The Ordovician is not by any means so finely calibrated.

Even where graptolite-bearing rocks are not exposed at the surface, the graptolites may prove their worth in correlating boreholes drilled to examine the subsurface geology. An example occurred many years ago when some boreholes were drilled in central England in search of coal. The rocks looked quite promising and not unlike some parts of the Coal Measures. But the discovery of graptolites in the borehole cores was enough to stop the drilling. Their presence indicated that the rocks being drilled were far too old to contain coal. As drilling is an expensive operation, a lot of money was saved in this instance.

A second example comes from eastern Europe where in the Baltic states of Estonia and Lithuania and in Southern Poland there has been extensive and detailed mapping of the subsurface Lower Palaeozoic rocks. This has been achieved by drilling numerous boreholes and correlating from core to core on the basis of their fossils, especially the graptolites.

A third example is of the Bendigo goldfields in Australia. Here, mining engineers identified rock strata on the sequences of graptolite shapes they found, at a time when academics had not yet sorted out the sequence of graptolite shapes! This was a practical application *par excellence*, and the gold-rich quartz sandstone 'reefs' could be precisely located.

There are inevitably problems in using graptolites for correlation. During the Ordovician in particular, graptolites were less cosmopolitan in their distribution, tending to be restricted to faunal provinces governed by geography and climate. In these circumstances, correlation may be straightforward between rocks within a single faunal province, but between different provinces it is generally more difficult unless there is a mixing of faunas to provide clues. Correlation using graptolites will, of course, depend on accurate identification of the species. In identifying graptolites, much importance is placed on various measurements (see Ch. 11). Unfortunately, graptolites are often found in rocks which have been contorted by earth movements which may result in lengthening or shortening of specimens and the distortion of various measurable features. The biostratigrapher may allow for a certain amount of distortion when identifying graptolites; but too much and the material becomes useless.

The usefulness of graptolites was not always appreciated. In the early days of geology, they had been consigned to a palaeontological backwater, while their bigger, more impressive and less perplexing contemporaries such as the trilobites, brachiopods and corals got all the attention. A single piece of work, stunning in both its brilliance and its effect on the geological world, pulled them out of obscurity.

The scene: the Southern Uplands of Scotland, a range of hills constructed of rocks which are monotonous in their sedimentary characters, but highly complex structurally, the strata having been severely crumpled and dislocated by ancient earth movements. This combination of characters can cause – and did cause – huge problems in interpretation, for without a means of establishing which strata are related to which, it is more or less impossible to 'unravel' the contorted and dislocated strata to arrive at a true understanding of the structure.

In the mid-nineteenth century, such eminent geologists as Nicol, Carrick Moore, Sedgwick and Harkness appreciated the broad structural trends and the basic stratigraphical relationships of the rocks of the Southern Uplands. But they made no progress in understanding the detailed structure because they saw no way of working out how the contorted steeply dipping rocks in each outcrop were related to each other. In particular, it was unclear whether the rock sequence observed was essentially unbroken and of huge thickness, or whether it was a thinner succession of rocks which simply repeated itself at the surface by the effect of folding or faulting.

Charles Lapworth (1842–1920, see p. 143) was a schoolmaster in Galashiels. For the sake of his health he used to walk in the nearby hills and he became interested in their geology. On a visit to Dobs Linn near Moffat he had an insight which was to revolutionise thinking about the Southern Uplands. Graptolites were to play the crucial role.

The key to Lapworth's success, where his predecessors had failed, lay in his recognition of the remarkable and distinctive changes that took place in graptolite populations through geological time. He observed that the graptolites occurred in distinct associations. By establishing a sequence of associations, or *zones*, Lapworth was able to correlate rocks of similar age over large areas. In this way he conclusively showed that the monotonous mudstones and sandstones that make up the Southern Uplands consist of numerous structural repetitions of a relatively thin sequence.

The importance attached to graptolites prior to Lapworth's classic work was aptly summed up by Lapworth's himself. He described the graptolites as 'A totally distinct group of fossils, and one hitherto regarded as of little geological significance' (1878, p. 242). The scale of Lapworth's achievement is difficult to describe: only James Hall (1865) had previously considered the real value of graptolites. Lapworth's demonstration of the detailed usefulness of graptolites as stratigraphical tools for the first time was a quantum leap in the understanding of Lower Palaeozoic sequences world-wide. In complex ground, and without pre-existing work to draw upon, he devised a thesis that not only removed all the difficulties of the orthodox theories for the structure of the Southern Uplands, but also established the value, previously quite unexpected, of graptolites in Lower Palaeozoic stratigraphy.

The zones established by Lapworth for the Ordovician and Silurian rocks of the Southern Uplands remain in use almost unchanged to the present day. Furthermore, the section chosen by Lapworth as 'typical' remains the type section for Lapworth's zones, and has recently been selected as the international stratotype section for the Ordovician-Silurian boundary. Though the structural interpretation of the Southern Uplands has been superseded, most, if not all of Lapworth's stratigraphical conclusions have stood the test of time and are still used in modern structural syntheses of this most complex area. Lapworth himself went on to become Professor of Geology at Birmingham University. His remarkable achievement in the Southern Uplands is commemorated by a plaque on the wall of Birkhill Cottage, near Dob's Linn.

To sum up, graptolites can be wonderfully useful in helping us to understand the underground architecture of large areas of the world. But they are not always easy to work with, especially when they are poorly preserved. So to finish this section with a cautionary tale: of a pencil-line drawn on a rock by a band of bored if inventive students on a particularly unproductive, miserable and rainy university field-trip. This was handed, with heavily applied sincerity, to an editor of this volume who shall remain nameless, as a major palaeontological discovery. 'Perhaps', said the sage who took it back to his lab. where he identified it as graphite from a 2B pencil in the space of ten minutes. But the damage had been done. The students were extremely happy to think they had gulled an expert in the field, and they spread the story far and wide. As far, indeed, as the magazine 'Private Eye' where it was magnified most colourfully into a collapse of a part of the Edifice of Science. Moral: even when you study the best behaved of fossils, these things happen if humans get involved.

Chapter Fourteen

WHY THE CONTROVERSIES?

Perhaps it is the inward-looking viewpoint of graptolithologists that gives the impression that the subject is perpetually involved in controversies. Perhaps all science is the same! It cannot be denied, however, that in the eighteenth and nineteenth centuries the preparatory techniques were relatively primitive and that this resulted in bizarre suggestions as to the affinities of graptolites: in essence there was widespread misinterpretation of their morphology. Thus one geologist considered graptolites to be the serrated tail spines of a primitive species of ray! Others considered them plants, cephalopods or bryozoans. The early debates, then, concerned the affinities of the class, and these were not resolved to any extent until the work of Kozlowski and Bulman in the immediately pre and post-Second World War period. One of the most convincing opposing viewpoints on affinity was that of the Swedish zoologist Berger Bohlin (1950) who deduced an extrathecal membrane and suggested affinity with the coelenterates. This argument has been continued and extended in modern times by two workers in particular (Dennis Bates and Nancy Kirk e.g. 1984), and it is a position not without merit as indicated earlier on in this book.

More recently controversy has hinged on the mode of life with the interpretations of, on the one hand Nancy Kirk and Dennis Bates (1984) and on the other Oliver Bulman (1964) and Barrie Rickards (1975). It may be that some of these problems are being resolved at present (e.g. Susan Rigby and Barrie Rickards, 1990 Chapter 5) yet others, such as the secretory mode *vis à vis* nemata, may be further from the solution (e.g. Charles Mitchell and K. J. Carle, 1986, Dennis Bates, 1986).

Always there is biostratigraphical debate about graptolites. This merely reflects the inverse ratio of the number of experts to the number of biostratigraphic problems needing solution: there simply is not the number of researchers to go round! (in the 1980s there were possibly 150 graptolite workers on a world-wide basis: only China and the Soviet Union could claim to have enough to deal with the problems surfacing, let alone to cope with the ordinary run of the mill service identification work).

As a result, there have been some 'classic' debates in the literature, such as that concerning the (Ordovician) correlation of the *Didymograptus bifidus* zone. In this instance the debate hinged on whether the American and European forms were both referable to the same species. The importance was, quite simply, that the nature of the widely-recognised Atlantic and Pacific provinces rested upon the correlation: if *bifidus* (American) = *bifidus* (European), then one kind of interpretation was possible; if *bifidus* ≠ *bifidus*, then a different interpretation was necessary. It was ultimately established, after great discussions in the literature taking up several years, that the resemblance between the American and European forms was purely superficial, based largely upon silhouette outline. As we have emphasized in this book, what you see of a graptolite in the rock, on the bedding plane, is not necessarily what you see when the specimen is totally isolated from the rock. To be fair it should be noted that graptolite zonation rarely hinges upon one species, and in this example suspicions were cast quite early upon the 'species' *D. bifidus* because of its associates: you can always tell a bad egg by the company it keeps!

So many of the controversies depend upon careful identification down to the specific level. When the fieldwork and specimens have not been carefully checked by the protagonists in debate then innumerable red herrings are raised before somebody solves the problem by revising the systematics. Systematic palaeontology (see Chs 11 and 12 and Appendix 6) is at the basis of all graptolite studies, however global in aspect, and it is good systematic palaeontology that will ultimately solve the problems.

A more recent debate than the *bifidus* problem has been the dating of the earliest land plants on Earth. One of the earliest land floras in Australia is the lower Devonian *Baragwanathia* flora of Victoria. These plants occur with undoubtedly Devonian monograptids at several localities. However, when Garratt was mapping for the Department of Mines he deduced that there were two plant horizons with *Baragwanathia*, one very much lower down the sequence. He eventually obtained graptolites from several of the localities with plants, including *Baragwanathia*; and he also established a sequence between the lower and higher floras, with graptolites between the two together with varied plant remains. It was concluded (e.g. Michael Garratt and Barrie Rickards, 1983) that the lower flora was actually Silurian (Ludlow) in age and that the whole sequence also spanned the Přídolí Series before the upper flora was reached. The conclusion seemed to cause a furore amongst some botanists and palaeobotanists, who clearly considered that *Baragwanathia* was too advanced a form to occur so early as the Silurian.

Nevertheless that does seem to be the position and the total evidence deduced by Garratt and Rickards seems overwhelmingly to indicate a Ludlow age for the earliest *Baragwanathia* plants: there is a remaining disagreement amongst graptolithologists about the identity of one species from one locality, but this hardly influences the information from the numerous other localities.

As a final example we can mention the bizarre case of a curiously named fossil *Promissum pulcrum* (meaning 'beautiful promise') described from South Africa as the earliest land plant (Eva Kovacs-Endrödy, 1987). Graptolite workers only became involved in this debate because some researchers considered these so-called plants more like graptolites than plants. There, is indeed, a superficial resemblance of their saw-toothed branches to Silurian monograptids, but on examination of the specimens the resemblance is clearly very superficial. Barrie Rickards concluded that the plants were, in fact, unusually elongate conodont elements in bedding plane assemblages, and they have in fact been described as such (Hannes Theron, Barrie Rickards, and Richard Aldridge, 1990).

Thus the controversies involving graptolites may be about the animals themselves, their evolution, mode of life, affinities and so on; or they may concern the relative age dating of rocks by means of graptolite assemblages and involve other disciplines. In any event controversies will not go away: they are part of the subject, and one of the attractions of the subject. Old controversies may re-surface to be re-opened as greater knowledge and refinement of techniques causes us to re-examine established tenets, things which seemed like milestones of discovery at the time. In the end everything comes down to collecting specimens with care, processing them with the latest techniques, and establishing a firm systematic base to the subject.

Appendix 1

WHAT DOES THAT TERM MEAN?

In this section we have included all the terms referred to in this book, and in addition other terms encountered in some of the papers listed in the references to the various chapters. Where terms are more or less obsolete we have noted them in italics.

acritarch(s) – Greek 'uncertain origin'. Minute (7–100 μm, generally < 150 μm) hollow organic walled bodies of very variable shape and sculpture. Found as residues from the chemical dissolution of a wide variety of marine sedimentary rocks from late Precambrian to the present. This artificial group was devised as a utilitarian one to encompass all such organic walled microfossils of unknown or uncertain origin and it is likely that it includes the various resting stages of unicellular algae, some plant spores, eggs of various invertebrates (including those of graptolites?), dinoflagellate cysts etc. They are often extremely abundant as fossils, especially within nonbioturbated mudrocks and probably include a significant proportion of the preservable phytoplankton, especially in the Lower Palaeozoic where they have considerable biostratigraphical potential. Group Acritarcha – the classification is an artificial, non-Linnaean one based purely on morphology without regard to relationship. Late Precambrian, extant.

anastomosis. Fusion, as of adjacent branches to form an ovoid mesh.

ancora. A branching prolongation of the virgella, present in Silurian retiolitids and some petalograptids.

anoxic. Literally without oxygen. Synonymous with *anaerobic*: without air, a common condition of stagnant or restricted circulation, bottom waters and the sediments deposited in them, especially in deep lakes and silled marine basins. The characteristic sediments are muds, often black and sulphidic because of the activities of sulphide reducing bacteria which generally inhibit the development of other benthic life forms (see *pyrite*).

apertural spine. A projection originating on the margin of a thecal aperture.

appendix. Narrow, distal portion of retiolitid rhabdosome, extending beyond the thecae (Fig. 67).

aseptate. Biserial rhabdosome lacking median septum, proximally (Figs 21–22).

asexual (reproduction). Common feature of colonial organisms whereby multiplication of the colony can be achieved without recourse to the hazards of fertilisation and larval development. In graptolites asexual reproduction is by budding of successive daughter individual zooids that are genetic replicas, i.e. a clone.

astogeny. The growth history of a colonial animal.

auriculate. Expanded, ear-like lateral lobes in highly modified thecae.

Autocortex. Cortical envelope of an individual theca, especially of an autotheca.

automobile. Literally self-moving, alluding to the postulated ability of graptoloid rhabdosomes (albeit very limited) to move through the water column by co-ordinated beating of the zooid tentacles.

autothecae. Larger type of regularly-developed graptolite thecae (Fig. 16).

axil. Base of V-shaped bifurcation of dichotomously branched rhabdosomes (Figs 83–84).

axonolipous. Graptoloid rhabdosomes which are not scandent and therefore do not enclose nema (Fig. 18).

axonophorous. Scandent biserial and uniserial graptoloids in which nema (or virgula) is enclosed within rhabdosome or embedded in dorsal wall (Fig. 22).

bandages. Ribbon-shaped components of periderm, laid down over the fuselli, usually on the outside of the thecae (Fig. 19).

basal disc. Discoidal plate developed from apex of sicula for attachment of some sessile graptolites (Figs 26 and 46).

bedding plane. A surface parallel to an original substrate of deposition. Generally only visible when marked by a change in deposition, e.g. of grain size, sediment composition, colour, etc. Where the discontinuity is quite marked the rock tends to part or break readily along that surface.

benthic – adjective; **benthos** – noun. Those organisms living at or near the sediment-water interface Fig. 125. Includes largely sedentary and fixed forms such as most dendroids, brachiopods and bryozoans, mobile forms that are bottom dwelling, e.g. flat fish, those with limited mobility, e.g. bivalves, and those that burrow or bore into the substrate, e.g. infaunal worms.

biform. Graptoloid rhabdosome (especially monograptids) with proximal and distal thecae of conspicuously different form (Fig. 110).

biostratigraphy. The study of sedimentary strata by means of their contained fossils based on principles established early in the nineteenth century, which claim that the same strata are always found in the same order of superposition and contain the same particular fossils. By identifying and matching the succession of fossil assemblages contained within separate sequences of strata correlation can be achieved between them. However, since many, if not most organisms, have limited geographic distributions (*endemic*), being controlled by substrate type, water temperature, larval dispersion etc., they are often restricted to particular types of sediment (*facies fossils*) and can only be used for correlation on a local scale. For more extended correlation, more wide ranging (*cosmopolitan*) fossils are needed that are not confined to particular facies (*facies breakers*). Thus organisms that are free floating or drifting (*planktonic*) forms or swimming (*nektonic*) ones should be the most suitable for correlation but three other major criteria are required. They should evolve rapidly, otherwise the time units they span would be too long; they should be abundant and they should be readily fossilised. Graptolites fulfil all these criteria quite well and hence are most important for the correlation and biostratigraphy of Ordovician, Silurian and some Lower Devonian strata. Also various microfossils, e.g. conodonts, ostracods, plant spores etc. are becoming increasingly important and providing complementary zonal schemes. The fundamental biostratigraphic unit based on such fossils is called a *biozone*.

bioturbation. Disturbance of a sediment by organisms (especially infaunal ones, Fig. 138). This may include activities such as feeding, burrowing, locomotion, even resting marks may disturb fine grained sediment and leave a preservable trace. We can learn much of the habits of fossils by studying the preserved traces of such activities within sedimentary rocks. Unfortunately intense activity completely destroys the traces and effectively homogenises the sediment in the way that earthworms help produce a soil.

bipolar. Monograptid rhabdosome with sicular cladium or pseudocladium.

biserial. Scandent graptoloid rhabdosome with two series of thecae enclosing nema (or virgula) (Fig. 22).

bitheca. Smaller type of regularly-developed graptolite thecae (Fig. 16).

bivalve(s) (synonyms – pelecypods, clams, lamellibranchs). The familiar and enormously successful (over 10,000 living species) and diverse group of benthic molluscs with two lateral mineralised shells (calcium carbonate, calcite and/or aragonite) enclosing the body soft parts. They have dominated marine shelf shelly faunas since the Trias. From an origin in the Cambrian they were initially eclipsed and marginalised by the brachiopods. But their greater adaptability and ability to burrow into sediment has allowed them to occupy a much greater range of ecological niches than those held by the brachiopods and also to invade fresh waters (from the Devonian). They are generally not found in association with graptolites except in certain relatively deep water marginal shelf edge environments with low oxygen levels and fine grained sediments that were generally unsuitable for brachiopods. Phylum Mollusca, Class Bivalvia, Cambrian, extant.

blastocrypt. Encysted internal layer of electron-dense crassal fabric, permeated by canaliculi, in graptoblasts.

blastotheca. Outer totally sealed peridermal layer of graptoblast, composed of fuselli, and continuing from stolothecal fuselli of crustoid.

brachiopod(s). 'Lamp-shells'. A remarkable group of shelled sessile invertebrates (5–200 mm generally < 30 mm, Fig. 125) that were dominant members of the shelly epibenthos of Palaeozoic shelf seas. Their importance waned through the Mesozoic and Tertiary to the present and they now constitute a very minor remnant as they were increasingly driven into marginal niches by the increasing success of the molluscs, especially the bivalves and gastropods. They are characterised by two external hinged and mineralised valves (positioned dorsally and ventrally) enclosing the body organs. They function like a simple filter pump using a filamentous lophophore for feeding and respiration. They also have a fleshy stalk extended through the hinged (posterior) end and generally used for attachment though atrophied in some. The group is divided into two classes, the Articulata and Inarticulata. The former, as the name suggests, has the shells hinged by teeth and sockets and mineralised with calcium carbonate (calcite). The latter do not have a hinging mechanism but instead a complex musculature for moving the shells relative to one another and mineralisation is generally by calcium phosphate with a high organic component of chitin. The high point of inarticulate diversity was in the Lower Palaeozoic when their ecological niches included muddy environments with low O_2 levels and little benthos, hence their association with graptolites. Phylum Brachiopoda, Lower Cambrian extant.

bryozoan(s) – 'moss animals'. An interesting and remarkably successful group of skeletonised small (1 mm – 10 cm, rarely up to 1 m) aquatic and sessile epifaunal invertebrates that form colonies (from 10^1 to 10^7 individuals) of varied shape. They are predominantly marine with a few fresh water species, the former having been important members of the shelly shelf faunas of the past, whilst the latter have no known fossil record. Biologically and morphologically there are similarities between the bryozoans, the dendroid graptolites and the living pterobranch hemichordates in that they all form colonies of quite closely integrated zooids with ciliated tentacular feeding organs called lophophores housed within an exoskeleton. This is generally mineralised in the bryozoans with calcium carbonate. Phylum Bryozoa; Ordovician, extant.

canaliculi. System of varied canals penetrating much of the blastocrypt in graptoblasts, perhaps representing voids between packed fibrils.

cauda. Narrow, proximal portion of prosicula (Figs 47 and 48).

central disc. Web of sclerotized tissue uniting proximal ends of stipes in certain graptolites (Fig. 112).

cephalopod(s). A large and diverse group of active predacious marine carnivorous molluscs that includes the familiar octopus, squid etc. and in the past many shelled forms such as the ammonites and their Palaeozoic ancestors, of which the pearly *Nautilus* is the sole surviving descendant. They are the largest and most abundant marine invertebrates and include many agile, rapid swimmers with shoaling habits. The calcium carbonate (aragonite) shelled forms diversified spectacularly in the Ordovician with a range of orthoconic (literally straight cones) forms whilst most of their descendants were coiled. These ranged in size from 5 cm to 10 m (generally $<$ 50 cm) and mostly had straight, gradually tapering conical shells with the squid-like animal living in an apertural body chamber. It is likely that their habits ranged from being bottom dwelling, sluggish swimmers to fast swimming, ocean going forms with shoaling habits. The nektonic habits of the latter, their abundance and rapid evolution give them great biostratigraphical potential which has been well fulfilled by the Mesozoic ammonites but because of the lack of well preserved features in many of the Lower Palaeozoic orthocones, the potential has not been able to be realised in the same way. Indeed Lapworth referred to graptolites as the 'ammonites' of the Lower Palaeozoic because of their potential use as zonal indices. Phylum Mollusca, Class Cephalopoda. Upper Cambrian, extant.

chert (flint). A hard, often brittle and glassy sedimentary rock consisting of microscopic quartz crystals (SiO_2) and sometimes amorphous silica. Occurs mainly as nodules in limestones and dolomites but may also form more extensive bedded deposits. It may be an original precipitate at or near the sediment-water interface or a replacement product. Often contains fossils, especially siliceous radiolaria and sponges but also calcareous echinoids and organic graptolites and inarticulate brachiopods.

chitinozoan(s). Problematic flask shaped microscopic (50–200 μm) and organic walled fossils often found in 'aperture to base' connected chains. They can be etched from a wide variety of Lower Palaeozoic marine sedimentary rocks and often occur in great abundance (hundreds per gram of rock). Their biological affinities are not known but they have a broadly common morphology and are regarded as a monophyletic animal group. Their wide distribution indicates that they were part of the zooplankton and from their abundance clearly an important element at that. Kozlowski suggested that they may have been graptolite egg capsules because of their close association with graptolite bearing strata. Along with various other problematic organic walled microfossils included in the acritarchs, they have considerable biostratigraphic potential. Tremadoc, Lower Ordovician to late Devonian (i.e. similar to graptoloids).

clade. All taxa distal to a given node in a cladogram; also, a monophyletic taxon consisting of an ancestral species and all its known descendant species.
cladium. Stipe developed from sicular or thecal aperture (Fig. 58).
clathria. Skeletal framework of rods (lists) composing rhabdosome, often supporting attenuated periderm (Fig. 66).
cleavage. The property or tendency of a rock to split along secondary (non depositional) aligned planes or textures that have been produced by deformation or metamorphism.
climacograptid theca. Strongly geniculate or knee-shaped theca with straight or slightly convex supragenicular wall parallel to axis of rhabdosome and relatively short (deep) apertural excavation (Figs 100 and 101).
coelenterate(s). A large, and essentially, marine group characterised by two body layers separated by a gel-like middle layer (mesoglea), a generally radial symmetry and two predominant body forms polyps and medusae. Only a few groups have any preservable hard parts but amongst those that do are the corals which have considerable importance. Reproduction may be either sexual or asexual with both solitary and colonial forms. The group includes hydroids (Class Hydrozoa), jellyfish (Class Scyphozoa), sea anemones and corals (Class Anthozoa), also graptolites according to some zoologists and thus a wide range of benthic and planktonic forms. Despite their lack of a skeleton, there is a long fossil record of the soft bodied coelenterates stretching back to the late Precambrian and the famous Ediacara Fauna. Rarely the impressions of planktonic forms have been found in black graptolite shales. Phylum Cnidaria, Class Hydrozoa, Class Scyphozoa, Class Anthozoa. Late Precambrian – extant.
coenoecium. Tubular exoskeleton of colonies of Pterobranchia (Fig. 33).
collagen fibrils. (see fibrils) (Figs 20 and 132).
colony. A variable concept in biology that includes both the social groupings (colonies) of insects and the much more morphological and physiologically integrated associations found in corals, bryozoans, graptolites etc. So in general a colony is a population of genetically similar organisms living in close proximity and interacting to some extent. However, more specifically within the marine invertebrates they are genetic clones founded by one individual (the siculozooid in the graptolites) with one or more generations, of asexually budded daughter individuals (zooids housed in the thecae), that have not separated from each other. Some genetic and morphological variation does occur such as in the thecal dimorphism of the dendroids within their specialised auto- and bithecae which are considered to have housed female and male zooids respectively. Graptolite colonies often show a sequence of morphological changes so that earlier individual the case may be different from later ones. Such changes of the colony as a whole are referred to as astogenetic and their evolutionary implications have been explored in some detail. Graptolites show a considerable degree of integration of their colonies (especially within the dendroids) and therefore would seem to have rather interdependent individual zooids. However, some monograptids are known to have been capable of regenerating broken stipes, even from distal fragments, showing that, here at least, groups of individual zooids were independent.
common canal. Continuous tubular cavity collectively formed by prothecae of graptoloid; rarely involving some portion of metatheca (Fig. 103).
conodont(s). An important and intriguing extinct group of small marine animals whose zooological affinities are only now beginning to be understood. Their mineralised

(calcium phosphate, apatite) elements can be etched, often in considerable numbers from carbonate shelf sediments of Cambrian, Ordovician to end Trias age. These elements are tiny (200 μm – 6 mm), toothed structures occasionally found in natural assemblages as a bilaterally symmetrical apparatus. Very recent discoveries of traces of their bodies seem to indicate that the animals were elongate (4–10 cm, very rarely up to 100 cm), eel-shaped swimming protochordates (and therefore distantly related to graptolites). The denticulate apparatus was probably an anteriorly placed jaw like structure for capturing their prey. Their distribution and abundance shows that there were both endemic (possibly benthic swimmers) and more cosmopolitan forms that lived mainly within shelf seas. Consequently they are proving to be very effective for biostratigraphic correlation and zonation, especially since they complement and extend the graptolite based zonal schemes into shelf environments. Some 140 conodont zones have been recognised from Cambrian to end Trias. ?Phylum Protochordata, Class Conodontophorida.

conus. Conical, distal portion of prosicula (Figs 47 and 48).

coral(s). This familiar and extensive group of marine animals have been important contributors to the sedimentary record of the earth since Silurian times because of their ability to build massive calcium carbonate reefs. Also, because of their high preservation potential they are an extremely important and much studied group of fossils. They are particularly useful in environmental reconstructions because of the limiting factors to their geographic distribution, e.g. water temperature and depth. However, generally, they are predominantly endemic forms and only of local or regional use in biostratigraphy. Some dendroid graptolites may have occupied niches within the quieter waters and fine grained muds flanking reefs and coralliferous mud mounds. Phylum Coelenterate, Class Anthozoa; Ordovician, extant.

cortical fabric. The most characteristic component of the cortical tissue, essentially comprising relatively robust, tightly packed collagen fibrils arranged in linear fashion, often in bundles.

cortical fibrils. (see fibrils).

cortical layer. Outer layer of graptolite periderm, composed of bandages.

cortical tissue. The component of the cortex as seen by light microscope studies (technically this could include fibrils). Under the electron microscope several *fabrics* are seen to occur.

cosmopolitan. An organism that has a very wide (global) distribution that crosses various geographical and ecological provinces.

crassal fabric. Electron-dense fabric occuring in various positions within graptolite peridermal tissue including stolons, the nema and retiolitid lists.

crinoid(s) – sea lilies. As their common name suggests, these solely marine animals are elegant and superficially plant-like echinoderms that generally lived attached to a substrate (epibenthic), rooted by long thin stems (20 mm – 10 m, usually < 1 m) and topped by an arm-bearing cup (calyx), although modern crinoids are often stemless and free-living. The arms are articulated and fan out to form an effective filter for feeding and respiration. There is a calcitic (calcium carbonate) endoskeleton of articulated columnals and plates that on death disarticulates (falls apart) into its constituent elements. These remains are often found with other fragments of brachiopods, bryozoans, trilobite exuvae, etc. as shell lags that have been winnowed and transported by bottom currents;

entire specimens are rare. Crinoids were a large and successful group during the Palaeozoic when they were the dominant echinoderms and lived in carbonate-rich shelf seas with other common reef dwelling organisms, e.g. bryozoans, corals, etc. Now they are the only surviving suspension feeding echinoderms and are dominated by their relatives, the echinoids and starfish. One Palaeozoic group (the scyphocrinitids) escaped the confinement of substrate dwelling by evolving inflated bulbous stem bases that probably buoyed them upside down in the plankton with the result that their remains may be found associated with graptolites in relatively deep water deposits. Phylum Echinodermata, Subphylum Crinozoa; mid Cambrian, early Ordovician, extant.

criss-cross fabric. See fusellar fabric.

crossing canal. Proximal portion of graptoloid theca which grows across axis of sicula to develop on side opposite that of its origin (Fig. 94).

crustacean(s). The most familiar modern members of this group, such as the crabs and lobsters, are but a very small part of what is one of the most diverse groups of organisms known. Originally marine, they have invaded almost all marine and freshwater environments and many terrestrial ones. They include the sessile suspension feeding barnacles (Class Cirrepedia), the scavenging and predacious shrimps and crabs (Class Malacostraca) and the small bivalved ostracods which themselves are very diverse and important to palaeontology and biostratigraphy. Typically they have segmented bodies (not ostracods) arranged in three distinct regions encased in chitinous based exoskeletons that are mineralised in some. Phylum Arthropoda, Super Class Crustacea; Cambrian, extant.

declined. Graptoloid rhabdosome with branches hanging below the sicula, subtending an angle less than 180° between their ventral sides (Fig. 87).

deflexed. Similar to declined but with distal extremities of stipes tending to horizontal (Fig. 96).

dendroid. (habit of growth). Bushy colony formed by irregular branching; and informal diminutive of Order Dendroidea (Figs 26 and 42).

denticulate. Sharply pointed thecal apertures bearing a short spine or mucro (Fig. 45).

diad budding. A division of the stolon, at a node, into two daughter stolons (Fig. 135).

diagenesis. All the chemical, physical and biological changes undergone by a sediment during lithification after deposition, exclusive of weathering and metamorphism. Thus embraces many different processes occurring at atmospheric pressures and temperatures, e.g. cementation, replacement, bacterial activity, leaching, formation of nodules, etc. These can radically transform the appearance of a sediment as it is changed into a rock.

diamond mesh. A mesh of criss-cross fibrils forming the outermost layer of the prosicula.

dicalycal theca. Graptoloid theca giving rise to two buds.

dichotomy. A branching point where both daughter branches diverge at the same angle from the direction of growth of the parent stipe (Fig. 83).

dinoflagellate(s). An extraordinarily varied group of minute unicellular algal-like organisms, characterised as their name indicates by two threadlike flagellae for locomotion. They are a very primitive and probably ancient group that cannot be simply assigned to either plant or animal kingdoms. However, the presence of chloroplasts is strong evidence of plant affinity. In habit they range from being parasitic to symbiotic (especially important here as the zooxanthellae symbionts of reef corals) to free living whilst the majority are planktonic. Today they are second only to diatoms as primary producers (i.e. photosynthetic phytoplankton) in both marine and freshwaters. As such they can be very abundant (up to 3000 per litre of water) and in certain conditions 'blooms' can occur (millions per litre), causing changes in the chemistry of the seawater leading to precipitation of calcium carbonate (as aragonite) and releasing toxins that can be fatal to other organisms. Many dinoflagellates produce distinctive organic walled cysts that are preservable and have been recognised with certainty as far back as the Silurian and possibly the Cambrian. Many fossil cysts currently within the acritarchs may well be dinoflagellate in origin. Their planktonic habit and abundance give biostratigraphical potential to the group and in the Silurian and from the Trias onwards they are increasingly useful for stratigraphic correlation along with the forams. Class Dinophyceae, Division Pyrrhophyta (20–350 μm). ?Cambrian, ?Ordovician, extant.

dipleural. Biserial graptoloid rhabdosome in which two stipes are in back-to-back contact so that each stipe has two external dorsal walls.

dissepiment. Strand of periderm serving to connect adjacent branches in dendroid rhabdosome; may be tubular (Fig. 44).

distal. Last-formed part (of stipe, theca, etc.) farthest away from point of origin.

dorsal. Side of stipe opposite thecal apertures.

echinoderm(s). A fascinating and diverse group of marine (largely stenohaline) animals that not only includes the familiar sea urchins (Subphylum Echinozoa) and sea-stars or starfish (Subphylum Asterozoa) but also the sea lilies (Subphylum Crinozoa), sea cucumbers (Class Holothuroidea) and a number of extinct fossil groups, e.g. the blastoids (Subphylum Blastozoa), homalozoans (Subphylum Homalozoa) etc. The phylum is characterised by the possession of hydraulic tube feet for locomotion and respiration and these are inter-connected by a water vascular system. They also have a virtually unique pentaradiate symmetry and most members have an exoskeleton of calcium carbonate (magnesian calcite) in the form of plates, spicules, spines, etc., hence the derivation of the name. So there is considerable preservation potential even if it is commonly as dissociated elements and these fossils are not uncommon in carbonate shelf-sediments. In the Cretaceous Chalk echinoids have been sufficiently common to be used as zonal indices and at the present similar burrowing echinoids are responsible for bioturbating sediment at all water depths. The phylum as a whole shows varied habits from burrowing (endobenthic) to attached (epibenthic) and active freeliving forms and inhabit all marine environments from rock pools to ocean depths. Some are herbivores (sea urchins), others filter feeders (crinoids) and yet other active predatory carnivores (starfish). All the major groups have long histories extending back to the Lower Palaeozoic and a few into the Cambrian. Phylum Echinodermata; Lower Cambrian, extant.

ectocortex. Component of cortex external to thecal tubes, composed at least largely of ectocortical bandages (Fig. 20).

endemic. An organism that is restricted to a particular region or environment.

endocortex. Component of cortex internal to thecal tubes, composed at least largely of endocortical bandages (often lacking bounding sheet fabric envelopes).

epibenthos. Those organisms living on or attached to the substrate compared with those which live within the substrate (endobenthos or infauna).

epicontinental. Situated on the continental shelf.

eucortex. Secondary deposit of cortical tissue with sheets, bundles of straight, parallel fibrils and ground substance (eucortex is herein considered as more or less synonymous with taeniocortex).

eurypterid(s). Sea scorpions. A bizarre extinct group that includes the largest arthropods known (2 m plus but more commonly 20 cm or so) and which attained considerable diversity during the Upper Silurian and Lower Devonian. They were vaguely lobster or scorpion-like with a segmented body and jointed appendages, including 6 pairs of legs with the posterior pair sometimes developed into paddles for swimming. The exoskeleton was chitinous and not particularly tough, so it is only well preserved where there were exceptional conditions for fossilisation. Their habitat ranged throughout the shelf seas into marginal marine, brackish and possible freshwaters, so as a group they were tolerant of a wide range of salinities (euryhaline). Most were active benthic forms but some were nektonic, especially the pterygotids with their swimming paddles and furthermore, as their spiny, grasping front legs indicate, they were active predators. Phylum Arthropoda, Class Merostomata, Order Eurypterida; Lower Ordovician-Middle Permian.

everted. Plane of aperture facing outward.

extensiform. Didymograptid with horizontal stipes (Fig. 62).

fibrils. Referred to as 'fibers' in old literature and subsequently as fibrils by other authors who considered the cortical (layer) fibrils to be probably collagen fibrils: Crowther and Rickards (1977) considered this proven and that fusellar (layer) fibrils were probably also collageneous; strictly speaking, can be used for any fibrillar material and can be either descriptive or, with prefix, genetic or conceptual.

flabellate. Rhabdosome fan-shaped (Fig. 42).

foramen. Aperture in sicular wall marking origin of initial theca.

foraminifera. An extraordinary group of minute (200 μ − 10 cm, generally < 1 mm) amoeba-like unicellular marine animals that secrete various kinds of skeleton (test) from purely organic to agglutinated with sand grains or other material to calcium carbonate. They can be sufficiently abundant to comprise a significant part of the sediment, e.g. the *Globigerina* ooze of the present deep ocean floor and the nummulitic limestones of the recent past, used to build the pyramids of Egypt. They are very diverse in their habits with many different benthic forms occupying all depths from the ocean floor to intertidal environments with fluctuating salinity (i.e. euryhaline). They have also formed an important constituent of the zooplankton since the Jurassic and are particularly important for biostratigraphy and especially for correlating recent ocean sediments and unravelling the recent history of the world oceans. Also, oxygen isotope studies of their tests allows determination of the temperature of the seawater in which they grew and thus are very important for palaeoenvironmental reconstruction. They have a long history extending back into the Cambrian, with the first chambered tests appearing in the Silurian, but the main diversification did not take place until the beginning of the Carboniferous. Subphylum Sarcodina, Class Rhizopodea, Order Foraminiferida; Lower Cambrian, extant.

free ventral wall. That part of the thecal ventral wall not forming the interthecal septum (Fig. 101).

fusellar fibrils. (see fibrils).

fusellar fabric. A peridermal fabric composed of criss-cross fibrils or fibrils arranged in a spongy texture.

fusellar layer. Inner layer of graptolite periderm, composed of fuselli (Fig. 19).

fusellar tissue. The components of the fuselli as seen by light microscope. Under the electron microscope several *fabric* types are seen to occur.

fusellus (pl. **fuselli**). A single growth increment of graptolite periderm, each one forming a ring or half-ring, arch-shaped in cross-section.

genicular spine. Sharp projection originating on geniculum.

geniculum. Angular bend in direction of growth of graptoloid theca (Fig. 100).

glauconite. A chemically complex green iron silicate mineral of the mica group $(K_1Na)\ (Al_1Fe^{+3},Mg)_2(Al_1Si)_4O_{10}(OH)_2$ that develops in marine sedimentary rocks with very slow rates of deposition. Can be found replacing microfossils, e.g. forams and ostracods and sometimes abundant enough to be a significant component of the sediment, e.g. glauconitic sandstones. Rarely found infilling graptolites.

glyptograptid theca. Sinuous theca with smooth curve in place of angular geniculum (Fig. 103).

granules. Informal descriptive term for small granules often occurring in membrane or sheet fabrics or close to same.

greywacke. An old term for dark sandstones, often poorly sorted with angular grains and rock fragments in a clay matrix. Largely confined to Palaeozoic marine basins and it is likely that much of the clay matrix is diagenetically derived from the contained feldspar grains and rock fragments. Sedimentologically they often have all the characteristics of *turbidites* being deposited in thick, laterally persistent sequences interbedded with shales (with graptolites in the Lower Palaeozoic) and sometimes submarine lavas and cherts.

ground substance. Possible protein/polysaccharides in physical support of straight, cortical fibrils; (interpreted by Crowther and Rickards (1977) as sectioned annular rings (annulations) on collagen, so the term becomes obsolete) (see also virgular fabric).

gymnocaulus. Unsclerotized stolon situated behind terminal bud in *Rhabdopleura*, from which zooids are proliferated.

half-ring. The form of most individual fusellae.

haematite. A common iron mineral (Fe_2O_3) of a metallic blood red colour that can be found as a primary constituent or as an alteration product in sedimentary, igneous and metamorphic rocks. Commonly infills graptolites.

hemipelagite. Sediment formed by very slow accumulation of biogenic and fine terrigenous material settling out of suspension. Often dark coloured and rich in organic material and characteristic of deep lakes with seasonal or little sediment supply and deep sea basins and their margins, typically rich in graptolites.

horizontal. Graptoloid rhabdosome with stipes disposed in plane at right angles to axis of sicula.

initial bud. Outgrowth through foramen or notch in sicular wall producing first theca of rhabdosome.

interthecal septum. Peridermal membrane separating overlapping thecal tubes in graptoloids (Fig. 115).

introverted. Plane of aperture facing inward.

isolation. Separation of distal portions of thecae from stipe as in, for example, *Rastrites* (Fig. 55).

lacinia. Delicate skeletal network, extraneous to rhabdosome proper (Fig. 114).

lacuna stage. Final period in development of porus in monograptids, where notch or sinus is closed by fusellar growth bands.
lappet. Broad, rounded, lateral apertural process of theca (or sicula).
limonite. A complex group of amorphous iron hydroxides commonly formed by oxidation (weathering) of iron minerals. Also formed by inorganic or biogenic precipitation.
list. Skeletal rod strengthening periderm in graptoloidea.
longitudinal rods. Stiffening rods found in fully-developed prosiculae, formed of closely parallel fibrils (Fig. 13).
lophophore. Paired arms or groups of arms, bearing tentacles and ciliated, situated adjacent to mouth of zooid; functionally food-collecting and respiratory (Fig. 9).
manubrium. The slightly bulbous structure produced by the initial downward growth of proximal thecae prior to the upward flexure of stipes in isograptids (Fig. 51).
median septum. Partition in biserial graptoloids separating two series of thecae (Fig. 101).
membrane fabric. (see sheet fabric)
mesial. Middle portion of free ventral wall (supragenicular wall) of theca.
metasicula. Distal portion of sicula composed of normal fusellar growth bands (Fig. 15).
metatheca. Distal portion of graptoloid theca (Fig. 72).
monopleural. Biserial graptoloid rhabdosome in which two stipes are in contact laterally; and facing in opposite directions.
microfuselli. Narrow fuselli, often lacking typical fusellar zig-zag suture, but wedge shaped. Ultrastructure consists of much electron-dense membrane fabric and reduced fusellar fabric.
monopodial growth. Type of colonial growth with permanent terminal zooid behind which new zooids arise as stem elongates.
multiramous. Branches numerous (Fig. 2).
nema (pl. **nemata**). Threadlike extension of apex of prosicula (Fig. 18).
nematularium. Modified, distal extension of nema.
neritic. That part of the sea and its substrate between low-tide level and the edge of the continental shelf (approximately 200 metres).
nodule. A term used to describe a wide variety of mineral masses found in a rock matrix and usually of contrasting composition to that matrix. Common sedimentary nodules are made of calcium carbonate, silica (chert or flint), pyrite, manganese and phosphate. Most appear to be of diagenetic secondary origin.
obverse. Aspect of graptoloid rhabdosome in which sicula is most completely visible (Fig. 23).
orders (of branching). Successive divisions of dichotomous branches, or successive generations of cladia.
ostracod(s). An ancient and important group of minute (0.5–30 mm, generally < 10 mm) bivalved crustaceans that have diversified from a marine origin in the Cambrian to occupy almost all possible aquatic niches. Their habits are equally diverse, ranging from benthic crawlers and burrowers to pelagic swimmers and planktonic floaters and some have even become parasitic. The bean-shaped shell (carapace) is generally mineralised with calcium carbonate (calcite) and hinged dorsally enclosing the body. Like most arthropods, the exoskeleton and carapace has to be regularly shed and newly secreted to allow for growth. This moulting enhances the preservation potential and they can be found in many different sedimentary facies, sometimes in sufficient

abundance to be rock forming, e.g. ostracod limestones and marls. The group as a whole has tremendous potential for correlation, environmental reconstruction and evolutionary studies. Some members of one group (the myodocopids) seem to have had cosmopolitan distributions since the Lower Palaeozoic, indicating a pelagic, if not planktonic, habit. They have been found associated with Silurian graptolites in black shales (Brittany), laminated muddy silts (Welsh Borders) and in shelly limestones (Sardinia, Bohemia) made predominantly of orthoconic cephalopods that were themselves nektonic. Phylum Arthropoda, Class Crustacea, Subclass Ostracoda. Upper Cambrian, extant.

palaeobiogeography. The study of the geographical distribution of faunas and floras in the past. The existence of regional patterns (provinces or faunal regions) in the distribution of life at the present has long been recognised. The possibility of there having been a single floral province during the late Carboniferous and Permian connecting South America, South Africa, India and Antarctica (called the Gondwanan province) was used as one of the main arguments in support of continental drift, that has since been validated by an understanding of plate tectonics. In the Lower Palaeozoic distinct shelly faunas have been recognised belonging to two faunal provinces on either side of a line running northeast-southwest approximately from north of Newcastle upon Tyne to Limerick in Ireland and continuing on the other side of the present Atlantic through Newfoundland and New England. This traces the Iapetus suture between the North American and European plates that moved together and closed this proto-Atlantic (Iapetus) ocean probably in the late Silurian. Even the graptoloids had phases when a significant number of species were endemic, resulting in the development of faunal provinces, e.g. during the lower Ordovician.

paracortex. Secondary deposit composed of multiple sheets with intersheet material of a condensed meshwork of fibrous material.

pectocaulus. Sclerotized stolon embedded in lower surface of mature parts of coenoecium of *Rhabdopleura* (Fig. 33).

pelagic. Referring to the water body as an environment rather than the bottom (benthic) or margins (littoral). Pelagic organisms are those that live in this environment and they may be swimmers (nektonic) or floaters (planktonic).

pellicle. That part of the membrane/sheet fabric forming the fusellar fabric envelope which overlaps with other similar parts to produce a layered structure (= pseudocortex, *pars*).

pendent. With approximately parallel branches hanging below sicula (Figs 89 and 94).

periderm. Material forming skeleton of Graptolithina, comprising inner (fusellar) layer and outer (cortical) layer.

plankton. Greek: 'wanderer'. Those passively floating or drifting organisms (mostly minute) inhabiting the water column and whose powers of locomotion are insignificant compared with the general movement of the water body. Thus distinct from these more powerfully swimming animals (nekton) and bottom living organisms (benthos), there is a basic subdivision of the plankton into the phyto − and zooplankton. The former (largely microscopic) are of fundamental importance in the food chain as primary producers capable of photosynthesising energy from sunlight, e.g. various algae, diatoms (siliceous), dinoflagellates (organic), coccoliths (calcium carbonate) etc., plus the larger blue-green and green algae. The latter zooplankton is taxonomically more diverse and often much larger, ranging from the microscopic protozoans (foraminifera, radiolaria, etc.) to invertebrates such as jellyfish, hydrozoans, annelids, cephalopods, some crinoids, arthropods, especially ostracods, etc. and even some fish. Also most importantly the microscopic planktonic larvae of a vast array of benthic invertebrates. Most of these zooplankton are the primary consumers of the food chain, feeding on the phytoplankton and each other. From the palaeontological point of view it is inevitable that only those plankton with mineralised skeletons or with tough organic walls are preserved in sediments but these include most of the groups mentioned above apart from the blue-green and green algae, the jellyfish, hydrozoans (and even these are occasionally fossilised) and the larvae of the invertebrates. As far as the Lower Palaeozoic is concerned there are some major groups of plankton that are now extinct, e.g. the graptolites, conodonts, chitinozoans and others that had not evolved, e.g. jawed bony fish and important algae such as the diatoms and coccoliths. Consequently there have been major changes in the composition and evolution of the plankton through time. All aspects of this major part of the global food chain are currently being investigated because of their importance to us all.

porus. Circular opening in wall of sicula through which initial bud passes to exterior (Fig. 15).

preoral lobe. Anterior glandular lobe or disc in pterobranchs, which secretes coenoecium.

primary deposits. Complement of the term secondary deposit; that material, largely fuselli, with which the rhabdosome is constructed prior to deposition of secondary material. The increments control the overall geometry of the colony.

prosicula. Proximal, initially formed part of sicula (Fig. 13).

protheca. Proximal portion of graptoloid theca (Fig. 72).

prothecal fold. Curvature of part of protheca (usually initial portion) giving noded appearance to dorsal margin of stipe) (Fig. 51).

proximal. First-formed portion (of stipe, theca etc.) nearest point of origin.

pseudocladium. Regenerated portion of bipolar rhabdosome.

pseudocortex. Secondary seposit of multiple sheets with scarce intersheet material devoid of fibrous material.

pseudovirgula. Virgula of thecal or sicular cladium.

pyrite. (Fool's gold). A very common iron sulphide (FeS_2) mineral with distinctive appearance similar to gold and often mistaken for it, hence the common name. It is commonly associated with fine grained marine sediments, especially in quiet waters where deposition is by slow settling of mud (predominantly clay minerals with high surface area to volume ratios) from suspension. Stagnation of the water reduces the diffusion of oxygen to a minimum especially below the sediment-water interface. Dead organisms falling into this environment produce a relative excess of decomposable organic matters over available oxygen. This leads to deoxygenation of the surrounding sediment and the growth of abundant anaerobic bacteria. Their active decomposition of organic sulphur compounds and reduction of dissolved sulphate produces hydrogen sulphide. This in turn reacts with iron containing minerals to form iron sulphide as the mineral pyrite. The crystals so formed may be microscopic and appear as minute spheres often concentrated at the sites of tissue decomposition, e.g. within a dead graptolite. (Fig. 29)

quadriserial. Scandent graptoloid rhabdosome composed of four rows of thecae in 'back-to-back' contact (Fig. 87).

radiolarian(s). These are remarkable free-living and predominantly planktonic microscopic (20–200 μm) amoeba-like protozoans. They secrete extraordinarily intricate skeletons of opaline silica, often with beautiful arrays of spines associated with complex, fine meshwork. At present their remains can be abundant enough to constitute a significant proportion of the sediment (radiolarian oozes) especially in open oceans lacking a supply of terrigenous sediments and this despite the fact that the majority of radiolarian skeletons suffer postmortem dissolution by seawater. Although they are an ancient group, they become progressively less abundant and less well-preserved in older sediments, partly because of dissolution but also because they are mainly associated with deep sea deposits and these become rarer because of subduction during movement of crustal plates. Nevertheless there are important surviving ancient remnants of the deep seas and these are often characterised by associations of submarine lavas (ophiolites), radiolarian cherts and black shales (with graptolites in the Lower Palaeozoic). Subphylum Sarcodina, Class Rhizopodea, Order Radiolaria. Cambrian, extant.

reclined. Graptoloid rhabdosome with branches growing upward, subtending an angle less than 180° between their dorsal sides (Fig. 88).

reflexed. Similar to reclined, but with distal extremities of the stipes tending to horizontal.

reticulum. Lists forming a distal prolongation of the ancora, thus producing a 'sleeve' around the rest of the rhabdosome.

retroverted. Thecal apertures facing proximally in consequence of hooked or reflex shape of metatheca (Fig. 72).

reverse. Aspect of graptoloid rhabdosome in which sicula is more or less concealed by crossing canals.

rhabdocortex. Cortical envelope of the whole rhabdosome or of a group of thecae. (NB. there has been a change in definition here, and the present authors consider that there are, anyway, problems involved in the usage of the terms autocortex and rhabdocortex. Autocortex has been used to mean that part of the cortex secreted on the outside of a young thecae before it was overgrown by succeeding thecae, when secretion of autocortex ceases. Autocortex is, therefore, thin and was for a while the outermost part of the growing colony. When the thecal growth has ended at any one point on the stipe, namely when the growing point has moved distally, then the outer cortical layer can thicken considerably. This is the layer which has been termed rhabdocortex. There is, however, no biological difference between the two, and we feel the terms are at best very arbitrary divisions in a continuous process. Further, it would be very difficult to apply such terminology where there is cortex both inside and outside the fusellar wall: the terms endo- and ectocortex seem far more satisfactory, and we suggest that rhabdocortex and autocortex be allowed to slip into disuse.)

rhabdosome. Sclerotized skeleton of entire graptolite colony.

root. Irregular branching structure developed from apex of sicula serving for attachment of some sessile dendroids.

scalariform. Preservational view presenting ventral (thecal) aspect of graptoloid rhabdosome, especially biserial forms (Fig. 30).

scandent. Graptoloid rhabdosome with stipes growing erect, enclosing or including nema (Fig. 22).

scopula. Bifurcating or ramifying structure branching directly off the nema to the obverse and reverse sides, of the rhabdosome.

seam. Ragged edge on list marking former attachment of periderm, common in retiolites.

secondary deposits. Any skeletal material laid down upon the (primary) thecal wall.

shale(s). A fine grained detrital sedimentary rock formed by the lithification (especially compression) of clay, silt or mud. Often finely laminated giving a fissility or parting parallel to the bedding along which the shale readily separates. May be various colours from white to black and generally associated with deposition in quiet waters. Bacterial activity may make the substrate inhospitable to most life with the result that the most common fossils found in shales are the remains of nektonic or planktonic organisms, e.g. graptolites. Metamorphism of a shale produces slate.

sheet fabric. Finely granular fabric forming the bounding layers of fuselli and cortical bandages.

sicula. Skeleton of initial zooid of colony, comprising conical prosicula and tubular distal metasicula (Fig. 17).

sinus stage. Initial phase in development of porus in monograptids, consisting of notch in apertural margin.

spiral line. Spiral arrangement of fibrils on the prosicula (Fig. 13).

sponge(s). An ancient group of biologically interesting and seemingly simple multicellular animals that are sufficiently different from most other metazoans to be placed in a separate subkingdom Parazoa. They are mostly marine sessile benthos (1 cm – several metres as reef-like masses), often skeletonised with spicules of either calcium carbonate, silica, organic spongin or combinations of these and consequently are important contributors to the sediment. They can be found in all water depths but are most abundant in certain shelf-environments often associated with corals and calcareous algae. The spicules of deeper water glass sponges are often associated with graptolites. Subkingdom Parazoa, Phylum Porifera. Cambrian, extant.

stipe. One branch of branched rhabdosome or entire colony of unbranched rhabdosome.

stolon. Electron dense, sclerotized sheath presumably surrounding unsclerotized thread of soft tissue, from which thecae appear to originate in Dendroidea and other groups (Fig. 34).

sympodial growth. Type of colonial growth in which each zooid is in turn terminal zooid of its branch.

synrhabdosome. Association of several graptoloids attached distally by their nemata to a common centre of unsclerotized material (Fig. 110).

theca. Sclerotized tube enclosing any zooid on rhabdosome (other than sicula) (Fig. 19).

thecal grouping. More or less regular association of groups of autothecae and bithecae forming small branches (twigs), particularly in acanthograptids.

triad budding. The division of the stolon, at a node, into three daughter stolons (Fig. 135).

triangulate theca. Type of isolate monograptid theca, triangular in lateral view, with hooked and horned apertures (Fig. 74).

trilobite(s). After dinosaurs, perhaps the most famous and familiar fossil group. They comprise an extinct class of Palaeozoic arthropods and thus are members of that most diverse and abundant phylum of invertebrates characterised by jointed appendages and including insects, spiders, crustaceans, etc. The trilobites were stenohaline, marine and mainly active benthic and consequently endemic animals of shelf seas, although some were pelagic and more cosmopolitan in their distribution. The segmented body has an exoskeleton of organic chitin mineralised with calcium carbonate that has to be regularly moulted to allow for growth. This ecdysis significantly enhances the fossil potential of the group but means that most trilobite remains are disarticulated. The group had its peak diversity during the Cambrian and Ordovician, often dominating the shelf shelly benthos of brachiopods, bryozoans and cephalopods. They can be found in almost all shelf facies from limestones, sandstones and shales, where they are occasionally found associated with graptolites. Because they were often very abundant and evolved rapidly, they are important for zonation and correlation. Phylum Arthropoda, Class Trilobita, Cambrian, extant.

turbidite. A sediment deposited from a turbidity current generally found in thick piles (hundreds or even thousands of metres thick) of repeated, well-defined sandstone beds (each 10 cm – 5 m thick) with considerable lateral continuity, even being traceable for several kilometres without much visible change. Each bed unit tends to be characterised by a vertical succession of sedimentary structures. There is a sharply defined base with sole marks (flutes, grooves, etc.) that have infilled previously eroded hollows in the underlying bed. Graded bedding is often evident with the larger, heavier grains at the base, sometimes including fossils. The top of the bed may have directional (current) ripples and grade imperceptibly into the overlying hemipelagic mud unit. These deposits are laid down by gravity powered density flows that are capable of covering considerable distances over very low angle slopes once the initial turbidity flow has been generated. Snow avalanches are terrestrial equivalents. At the present this type of deposit is found infilling ocean basins and yet are known to have been derived from the continental shelf. The initial flow is confined to the submarine valleys or canyons that cut deeply into the continental shelf slope then out into the ocean basins.

uniserial. Rhabdosome or stipe of graptoloid consisting of single row of thecae only (Fig. 74).

ventral. Side of stipe on which thecal apertures are situated or comparable side of thecal aperture.

vesicles. Ovoid or subspherical bodies, often in association with membrane fabrics, though located elsewhere occasionally, possibly representing sites of gas pockets or fat bodies.

virgella. Spine developed during growth of metasicula, part of sicular wall and projecting freely from its apertural margin (Fig. 50).

virgellarium. Umbrella-shaped structure developed at tip of virgella in linograptids.

virgula. Term commonly used for nema of scandent graptoloids (Fig. 22).

virgular fabric. A stellate-septate ultrastructure composed of virgular fibrils which appear lucent on a background of electron-dense matrix (as opposed to cortical fabric which has a matrix less dense than its fibrils). Virgular fibrils show an internal substructure whereas cortical fibrils are usually solid or have a lucent core.

zig-zag suture. Suture at junction of fusellar half-rings (Fig. 15).

zooid. Soft-bodied individual inhabiting theca or coenoecial tube (Figs. 8, 9, 33, 34).

APPENDIX 2

WHERE ARE THE GOOD PLACES TO COLLECT GRAPTOLITES?

There are thousands of graptolite-bearing localities in the British Isles, and many still await discovery. In the last decade alone the British Geological Survey, in the course of its mapping programme, collected graptolites from hundreds of localities, discovering many of them for the first time. So we suggest here only a very short list of localities, selected for ease of access and/or the richness of their faunas, and also to cover a wide range both geographically and stratigraphically.

If you go collecting at these or any other localities follow the advice in chapter 11 and remember:

Observe the Country Code;

Get the landowner's permission before you enter private land;

Follow the Code of Conduct devised by the Geologists' Association (obtainable from Geologist's Association, c/o Geological Society of London, Burlington House, Piccadilly, London W1V 0JU).

WALES

OGOF-DDU, 2 km east of Criccieth, Gwynnedd [SH 514 379] OS 1:50,000 map 124. The cliff exposure shows the base of the Tremadoc Series and yields the dendroid *Rhabdinopora* (Fig. 4). From the sea-front at Criccieth take the footpath eastwards alongside the railway (BEWARE of trains). About 0.8 km from Criccieth the lower Tremadoc forms a cliff on the N side of the railway. *Rhabdinopora flabelliformis* can be found at the place where the railway has been cut through a projecting crag of slate, and at several places in the fenced ground to the east and west. *Use your eyes* rather than your hammer! The rocks are strongly cleaved and you have to be sure to look for fossils along the bedding (shown up by faint colour-banding and sandy layers), not the cleavage-planes. *Rhabdinopora* helps your search because it shows up where it forms planes of weakness in the rock.

ABEREIDDI BAY, 7 km NE of St David's, Dyfed [SM 795310], OS Map no.157. At the south end of the bay dark slates of Llanvirn age are tipped up vertically and strike out to sea, forming large (and slippery) intertidal exposures (Fig. 137). Approach past Llanvirn-y-fran to the sea-edge and walk round (N.B. on a falling tide) to the south end of the bay. Where bedding and cleavage are parallel large *Didymograptus murchisoni* and diplograptids can be found strikingly preserved as whitish chloritic flattened impressions on the dark bedding-planes. Material in relief is rare; trilobites, gastropods and brachiopods also occur. The loose water-worn boulders often yield good specimens and are easy to split open. Watch the tide! If you are cut off do not attempt to scale the cliffs hereabouts as they are dangerous. It is safe to wait for the tide to go out again. Reference: Bassett, M. G. (ed.) 1982. *Geological Excursions in Dyfed, south west Wales*, pp. 51–63. National Museum of Wales, Cardiff.

TY'N-Y-FFORDD QUARRY, 7 km ENE from Llanrwst, Gwynedd-Clwyd border [SH 8699 6525], OS sheet 116. A mass of disturbed mudstone of the Elwy Group overlies well-bedded Nantglyn Flags. Both are of early Ludlow age, but the graptolites show that the disturbed beds, which have slid down into their present position in an underwater land slide, are slightly *older* than the strata that they rest on. The Nantglyn flags yield flattened monograptids with specimens of *Saetograptus, Neodiversograptus* and *Pristiograptus*. The Elwy Group yields fossil shells and graptolites (including *Monograptus*) in relief; but the beds are not very accessible without a step-ladder. Reference: *Geology of the country around Rhyl and Denbigh*, 1984. *Memoir of the British Geological Survey*. HMSO.

AFON SEIONT, Caernarfon, Gwynedd, North Wales. [SH 47886247 – 48066222], OS sheet 115. Dark grey sandy siltstone crops out below Caernarfon Castle, and though graptolites have been found here, *no collecting should be attempted*. The same rock crops out on the other (south) side of the river upstream from the boatyard. From the castle car park, cross the footbridge and turn left along the road to the turning into the boatyard: access to the river edge is through the boatyard. The shore can be muddy and is covered by water at high tides. Extensiform didymograptids and tetragraptids along with cyclopygid trilobites and phyllocarids can be recovered from the majority of outcrops, but their abundance varies. Beyond the wreck at the bend in the river outcrop deteriorates. The graptolites are preserved mainly as black periderm or silvery chlorite films, and some retain considerable relief.

LLANFAWR QUARRIES, Builth Inlier, on the eastern outskirts of Llandrindod Wells, Powys [NGR SO 065618], OS sheet 147. Exposures in the quarries are composed of fossiliferous, black shaly mudstones and a number of small igneous intrusions. Graptolites (and trilobites) can be easily collected from the mudstones, and there is commonly much loose, fossiliferous material lying around on the quarry floor. The graptolites are generally well preserved as flattened films of pyrite (iron sulphide). At least eight species belonging to the middle Ordovician *Nemagraptus gracilis* Zone are present. *Nemagraptus, Dicellograptus, Dicranograptus, Climacograptus, Cryptograptus*, and *Glossograptus* occur here. Llangurig A44 road section, 3 to 5 km WNW of Llangurig, Powys [SN 858820 to 864818], OS 1:50,000, sheet 136. There is convenient parking in a lay-by directly opposite the section at grid reference [861818]. This extensive section exposes upper Llandovery greywackes with dark layers which yield numerous graptolites. There appears to be a low angle of dip but this is in fact the plane of cleavage. The beds themselves are steeply dipping, and by standing back from the section the darker graptolite-bearing layers can be distinguished. The monograptids they contain are characteristic of the upper part of the *M. turriculatus* Zone, whilst graptolites indicative of the overlying *M. crispus* Zone occur about 2 km to the east at [875812; 879810]. The boundary between the two zones approximately coincides with the point at which a narrow road joins the A44 road at [863818].

CYFFENI WOOD, 1 km SE of Sarn Village, Powys [SO 204909], OS sheet 137. The entrance to the wood is along a minor road a little over 1 km east of Sarn village at [217909]. From there it takes only a few minutes to walk to the beginning of the track-side section at [217905], where a diverse assemblage of early Ludlow graptolites can be collected from dark grey silty shales. Saetograptids are especially well represented, and lobograptids and pristiograptids also occur. The section continues along the track to the WSW, slowly ascending the succession through the *N. nilssoni* and *L. scanicus* zones, until silty shales are succeeded by coarser sandy beds, which have been extensively quarried. Graptolites are not generally common at this higher level, but *P. tumescens* dominates, indicating the *P. tumescens* Zone.

THE RHEIDOL GORGE, 17 km E of Aberystwyth, Dyfed [SN 749800–749791], OS sheet 135. The River Rheidol here cuts through strata of Lower to Middle Llandovery age, predominantly black shales in the older rocks to the north but with an increasing proportion of turbidite siltstones and sandstones as the sequence is ascended southwards. Access is via a public footpath either from the George Borrow Hotel (W of Ponterwyd, SN 74678055) or from Yspyty Cynfyn (between Ponterwyd and Devil's Bridge [SN 75307905]. Graptolites abound, often superbly preserved as pyrite internal moulds. Diplograptids predominate in the lower beds with monograptids becoming common higher in the succession. Over 100 different species have been recorded. A traverse made from north to south collecting en route will indicate clearly the evolution of the Llandovery graptoloids after their near extinction at the end of the Ordovician, together with their considerable value as biostratigraphical indices.

ENGLAND

OUTERSIDE, 4 km SW of Keswick, Cumbria [NY 211215], OS Map 89. On the NW side of the hill, greyish slate of the latest Arenig to earliest Llanvirn is exposed, and forms screes overlooking Coledale Valley. Take the public footpath from Braithwaite to Sail and Grasmoor; after about 3 km, just before you get to the summit of Outerside, veer to the right to make for the screes. **Warning – do not venture there in icy or bad weather**. A variety of graptolites, mainly *Tetragraptus*, *Glyptograptus* and a range of *Didymograptus* species can be found on the screes, often as greenish flattened rhabdosomes. Some twenty species have been recorded here, together with crustacea, rare trilobites and brachiopods.

BARF, 6 km NW of Keswick, Cumbria [NY 218265], OS sheet 89. A large scree of Skiddaw Group (Loweswater Flats, mid-Arenig age) on the south-east flank of the mountain has been a popular collecting-ground for over a century. Approach by the bye road from Thornthwaite. From 'the Clerk' you can ascend the scree to the white-painted stone called 'the Bishop' looking for graptolites, some of which occur as shiny, flattened specimens and others as rusty impressions. About 15 species of graptolites have been found, including some *Pseudophyllograptus*, *Azygograptus* and slim species of *Didymograptus*. This scree has, however, been well searched in the past and graptolites are not as numerous as they once were. Exploring the less-frequented screes to the north or above the Bishop, is more rewarding, BUT as the face of Barf is very steep, take care and don't go there in bad or icy conditions.

SKELGILL, near Ambleside, Cumbria [NY 39640320] OS Sheet 90 or OS 1:50,000 Lake District Tourist Sheet. Parking for one car on the side of the Troutbeck-Holbeck Ghyll road near the entrance to the track leading north westwards to High Skelgill. The walk is about 1 mile from here. The site is protected and is well signposted before High Skelgill is reached. The Lower Bridge Section, to which the grid reference above refers, is detailed in Hutt, J. E. 1974 Palaeontographrical Society Monographs, 128, pt. 1, 1–56. Graptolites are abundant throughout the Llandovery, and good collections can be made from fallen blocks in the stream. Over 100 species recorded.

NEAR GILL, Cautley, north of Sedbergh, Cumbria [SD 97257050] OS Sheet 98. There is limited parking (one vehicle) at the point where Near Gill crosses the Sedbergh to Kirkby Stephen road. Proceed upstream on to open moorland: after 200 m of fairly steep climb the stream takes a sharp turn to the East, and a conspicuous bedding plane is seen in the left bank, dipping towards the stream. The best collecting at this locality is in the small screes at the foot of the conspicuous bedding plane and adjacent to nearby rock exposures. Numerous well preserved graptolites, silver on blue-grey background, can be collected from the middle Wenlock *linnarssoni* Zone of the Brathay Flags. There are several species of each of the genera *Pristiograptus*, *Monograptus*, *Monoclimacis*, *Cyrtograptus* and *Retiolites* s.l. Reference Rickards, R. B. 1967 *Quarterly Journal of the Geological Society of London*, 123, 215–251.

MOUTH of WANDALE BECK, Cautley, north of Sedbergh, Cumbria [SD 97757070] OS Sheet 98. Park a couple of hundred metres past the mouth of Wandale Beck into the R. Rawthey, where there is an official car park. Uppermost Llandovery rocks and lowest Wenlock (*centrifugus* to *riccartonensis* zones) are continuously exposed where the stream enters the R. Rawthey. Graptolites are often in full relief, preserved in golden pyrite, and are aesthetically pleasing. However, a heavy hammer is needed because the Brathay Flags are hard and tough. Good specimens of *Monograptus*, *Monoclimacis*, *Cyrtograptus* and *Retiolites geinitzianus* can be obtained quite readily. Reference: as for Near Gill locality.

WANDALE HILL, Cautley, north of Sedbergh, Cumbria, [SD 98227002] and [SD 989702] OS Sheet 98. Park in a lay by next to the R. Rawthey just short of Handley's Bridge to the left of Sedbergh-Kirkby Stephen road. Walk up the lane from Handley's Bridge to Narthwaite Farm. Immediately through the farmland take the *uphill* track, then take the left fork along the contour after 100 m or so. After a quarter of a mile this track reaches a quarry on the open fell, reached through a gate across the track. The quarry yields abundant Ludlow graptolites in three dimensions, including *Neodiversograptus*, *Bohemograptus*, *Pristiograptus*, and *Saetograptus*; also bedding planes covered in many flattened, but well preserved specimens of *Saetograptus incipiens*, occur in several parts of the quarry. Continuing northwards along the wall/track leads to several dry gullies running down from Wandale Hill just north of Mountain View Farm. These gullies yield a rich Upper Wenlock *ludensis* zone fauna from deeply weathered laminated mudstones. Fossils can be collected from the screes and in sites. Reference: as for Near Gill.

RIVER CLOUGH, east of Sedbergh, Cumbria [SD 69409150] OS Sheet 98. Park at the viewpoint car park on Longstow Common on the A684 Sedbergh to Hawes road [SD 694913]. A Sedgwick Trail locality guide can usually be obtained from an honesty box at the viewpoint (50p) or from the tourist office in Sedbergh. Follow the trail route in the *reverse* direction to that recommended, proceeding to localities 10, 11 and 12. Here Wenlock rocks, blue-grey laminated mudstones (Brathay Flags), red stained, are exposed beneath a red conglomerate which is unconformable upon them. Upper Wenlock graptolites are abundant in the mudstones, especially the genera *Pristiograptus*, *Monograptus* and *Cyrtograptus*. Collect carefully and do not leave debris obscuring outcrops. Graptolites can often be seen *in situ*, on the bedding planes, without hammering at all.

SCOTLAND

MORROCH BAY, 3 km SE of Portpatrick, Wigtownshire, Scotland [NX 014526] OS Sheet 82. Intertidal exposures of black shales, greywacke sandstones and cherts are of Llandeilo-Caradoc age. The bay is reached by the footpath at the S end. There is little to see at high tide, but at low tide there are many bands of Glenkiln and Lower Hartfell Shales, many of them fossiliferous, as described by Peach & Horne (1899, p. 401). The rocks are affected by faulting but with patience, rich faunas of the *Nemagraptus gracilis* to *Dicranograptus clingani* zones can be found, including species of *Leptograptus*, *Dicellograptus*, *Dicranograptus*, several diplograptids and *Corynoides*. Localities can be identified by pacing out distances to the igneous dykes on the foreshore. You can then compare results with those of Peach and Horne, J. 1899. 'The Silurian Rocks of Scotland'. *Memoirs of the Geological Survey*, UK, Scotland. HMSO.

HARTFELL SPA, 6 km N of Moffat, Dumfries and Galloway [NT 097 116], OS map 78. A large exposure of black shales is the type locality for the Hartfell Shales, of Caradoc age. Choose a day of mild or settled weather conditions for your visit. Hartfell Spa is reached from the small lay-by at NT 075 103 by taking the well-marked footpath that extends along Auchencat Burn for 5 km. The Spa itself is sheltered by a tiny stone hut on the south side of the stream. The strata on that side are more mangled than on the N side where the cliff is made up of hard black shales, mainly of the *C. wilsoni* to *P. linearis* Zones. The scree can yield fine slabs with *Leptograptus*, *Pleurograptus*, *Dicellograptus*, *Dicranograptus*, *Climacograptus* or *Orthograptus*, sometimes strikingly preserved as white flattened specimens. The curious *Corynoides* (usually 5–10 mm long and 1–2 mm wide) is present in thousands in some beds.

DOB'S LINN, 13 km NE of Moffat, Dumfries and Galloway [NX 196158], OS sheet 79. This is Lapworth's classic locality now an SSSI, where all the zones from *C. peltifer* (low Caradoc) to *R. maximus* (upper Llandovery) are present. Here also is the stratotype where the base of the Silurian System is now formally defined. From the lay-by SW of Birkhill Cottage (Lapworth's one- time lodgings) at [202158] walk down the side-stream into the valley; at the bottom turn right and go up the main stream into the gorge (N.B. it is dangerous in snowy or icy conditions). About 200 m into the gorge there are large cliffs of black shales; those on the south (right) are deformed but the 'Main Cliff' on the north has better ordered exposure of late Ordovician and early Silurian age. The scree yields good material, but you can also ascend the cliff to examine each bed in turn. Note that a part of the succession is in the grey Barren Mudstone and yields no graptolites, but the black beds below and above are fossiliferous. If you wish to know more about this locality, refer to the descriptions given by Lapworth (1878, 'The Moffat Series'.

Quarterly Journal of the Geological Society of London, 34, 240–346), Toghill (1968, Graptolite assemblages and zones of the Birkhill Shales (lower Silurian) at Dobb's Linn. *Palaeontology*, 11, 654–668) and Williams (1980, Geological Excursions to Dobb's Linn. *Glasgow Geological Society*).

MEIKLE ROSS, 10 km SSW of Kirkcudbright, Dumfries and Galloway Region, OS sheet 83. Meikle Ross is the peninsula which forms the southern extremity of the west side of Kirkcudbright Bay. Early Wenlock sediments are well exposed, forming reefs on both sides of the peninsula. They are best examined on the west side, even when the tide is high, and an additional advantage along this stretch of the coast is that the cliff is very low. Park near to Ross Farm [NX 646447], which looks out on to Ross Bay on the east side of Meikle Ross. Make your way to the west side of Meikle Ross either along the coast path around the headland or by a more direct westerly route across the fields (*first of all obtaining permission at the farm*). At a point where the coastline abruptly changes course from north-south to east-west (forming Fauldbog Bay) graptolite-bearing beds occur in the reef, believed to be of oldest Wenlock age (*Cyrtograptus centrifugus* Zone). Beds of the same age, but containing a more diverse assemblage of cyrtograptids, monoclimacids, monograptids and retiolitids occur some 300 m to the SSE at [644443]. Exposures of higher beds along the shore to the south contain graptolite horizons with the long slim and abundant *Monograptus riccartonensis* indicative of the *M. riccartonensis* Zone. Wenlock graptolites are found in thinly bedded or laminated silty shales, which have a characteristic brown speckled aspect. This lithology is in marked contrast to the dark graptolite-bearing shales of the underlying Llandovery Series.

IRELAND, NORTH AND SOUTH

COALPIT BAY, Donaghadee, County Down, Northern Ireland. [J 595788]. Dark shales of Late Ordovician and early to mid-Llandovery age are exposed, mainly between high and low water levels. The section is complicated by faults – see Carrickfergus Memoir (pp. 11, 20) for details. The Ordovician is exposed in a contorted area beside an igneous dyke and is difficult to investigate. The highest beds, adjoining the greywackes at the south end of the bay, have rich faunas of the *Monograptus convolutus* zone. Reference: Geology of the country around Carrickergus and Bangor (1982). *Memoir of the Geological Survey of Northern Ireland.* HMSO.

TOMGRANEY, near Raheen Bridge, County Clare, Ireland, [R 6483]. Limited parking is possible on the roadside near the locality. A small stream flows northwards through Raheen Bridge. Take a left turn on the Raheen Bridge – Tomgraney road some 200 yards west of Raheen Bridge. This road reaches another minor road after less than half a mile; turn left and stop after about 300 yds where the road crosses the stream. Just 100 yards upstream are small exposures in dark graptolite shale of *gracilis* zone (Ordovician, Caradoc) age. These yield numerous species of *Nemagraptus*, *Orthograptus*, *Dicellograptus*, *Dicranograptus*, *Climacograptus*, *Pseudoclimacograptus*, *Diplograptus*, *Amplexograptus*, *Lasiograptus*, *Hallograptus*, *Corynoides* and rare *Didymograptus superstes*. A tributary stream some 300 yards to the east yields well preserved Llandovery graptolites, especially the genera *Parakidograptus*, *Akidograptus*, *Climacograptus*, *Dimorphograptus*, *Cystograptus*, *Monograptus*, *Pristograptus*, and *Rastrites*. Reference: Rickards R. B. and Archer, J. B., 1969. *Scientific Proceedings of the Royal Dublin Society*, Series A, 3, 219–230.

BALBRIGGAN, County Dublin, Ireland. Park on the Skerries to Balbriggan road: there is a path leading from this road to the shore, about a quarter of a mile south of Fancourt, at the north end of a row of properties to the east of the road. Turning *left* reaches, after just over 200 yards, a series of gullies running through the exposed rocks of the foreshore. To the north of those gullies are Ordovician lavas and unfossiliferous brown shales: in the gullies are good exposures of black shales, much faulted, but yielding late Ordovician *anceps* zone graptolites and, close by, low Llandovery *acuminatus* zone faunas, and '*gregarius*' zone faunas. Numerous species can be collected. Turning *right* from where the road reaches the shore leads immediately to the Llandovery-Wenlock boundary, the upper horizon marked by the incoming of distinctly bluish flags. For the next few hundred yards these yield numerous Wenlock graptolites of the genera *Monograptus*, *Pristiograptus*, *Monoclimacis* and *Cyrtograptus*, though they may have become more difficult to collect up sequence. Reference: Rickards, R. B., Burns, V., and Archer, J. B., 1973. *Proceedings of the Royal Irish Academy*, Series b, 73, 303–316.

COLLECTING OUTSIDE THE UK AND IRELAND

Earlier in the book (Ch. 4) we pointed out that graptolites have been found on all continents except Antarctica. It is possible, using the principles outlined in Appendix 5, to visit museums and departments of geology in the Universities of large towns, and to find out where graptolites can be collected. Local small town museums can also help on occasion. But collecting in some countries, such as China, the USSR, and the Americas, may involve much greater commitments to travel than is ever necessary in the UK. The general locations which we list below are some which yield abundant graptolites once the right beds are found, and they are a reasonable distance from centres with normal accommodation facilities. Normal procedures with regard to collecting and permissions should always be followed, and if more than a few specimens are collected, permission may be needed to export them, should you wish to do so.

FRANCE

Accessible localities in France, which yield quite beautifully preserved Silurian graptolites, though not abundantly, are to be found near Rennes, to the north and north east. Close to Ardouillé-Neuville, some 300 m SE of the D.23/N776 intersection is a small quarry; and other workings 1 km NE of Médréac, both of which yield Llandovery graptolites in three dimensions. Other similar localities occur 1 km SW of Guitté close to the D.25, and 1 km SE of Vieux-Vy-sur-Coueson. Permission should be obtained to enter these sites. Guidance can be obtained from the Laboratoire de Paléontologie et de Stratigraphie at the Université de Rennes.

NORWAY

There are numerous rich graptolite localities a short drive from Oslo, as far away as Ringerike, and even close to the town itself. The fossils are often well preserved in shale and limestones, and in age they will usually be Ordovician and Silurian. There is an extensive literature on them (see how to get into the literature earlier in this Appendix) and help can be obtained from the Palaeontologisk Museum, Sars Gate 3, 0562 Oslo 5.

SWEDEN

Detailed information can be obtained from the Universities of Stockholm or Uppsala in the east, or the University of Lund in the far south. Very good localities occur very close to Lund, both of Ordovician and Silurian graptolites, and good comparative material is held by the University there. Northwest of Uppsala, the Siljan/Rättvik region is both very productive and pleasant to work in, but detailed references as to actual published localities are needed. Osmundberg and Silvberg are especially productive: permission can be obtained to look in the limestone quarries of the former; and the Silvberg locality.

SPAIN

The Catalonian coastal ranges near Barcelona, especially in the region of the Llobregat River a few km west of the town, have been widely studied in recent years, and graptolites obtained from numerous Silurian and Devonian horizons and localities. Preservation varies from good to tectonically deformed, and many of the localities are not easily accessible. Details can be obtained from the Department Geòtectonica, Universitat Autònoma de Barcelona. Other localities in Spain, and in Portugal, are very difficult to collect and are not recommended for casual visits.

CZECHOSLOVAKIA

The Barrandian area of Central Bohemia (Prague Basin) has long been famous for its fossiliferous Lower Palaeozoic strata and especially its Silurian-Devonian shelly faunas. However, graptolites can also be found and are of importance because of their use in correlation, particularly since the internationally recognised type section of the boundary between the two systems was established at Klonk (30 km southwest of Prague). Because of this international scientific importance a number of the most fossiliferous localities (often old quarries) are protected sites, e.g. the Pozary Quarries, 1–5 km east of Reporyje, on the southwestern edge of Prague.

Guidance to publicly accessible localities may be obtained from the Geological Survey of Czechoslovakia, Mala Strana, Prague 011 or the National Museum, Vaclavske namesti 68, Prague 1 – Stare Mesto. The latter has one of the best collections of Lower Palaeozoic fossils in the world, largely as a result of the work of Jaochim Barrande (see biographical sketches).

YUGOSLAVIA

Although generally Yugoslavian graptolites are badly preserved, there is considerable diversity of form, some 115 species being recorded from Zvonacka spa in Eastern Serbia where Silurian graptolites occur. Similar strata occur in the Rtanj Mountains 150 km SE of Belgrade. Advice should be obtained from the Natural History Museum, Njegoseva 51, Beograd.

SARDINIA

In southeastern Sardinia (the River Flumendosa flows through the hills of the Gerrei and Sarrabus regions (northeast of Cagliari) from the villages of Goni down to Villaputzu and the sea and for much of its course is followed by the road from Ballao to Villaputzu. Within this 30 km belt there are numerous hillside sections in Silurian black graptolitic shales ranging from Middle Wenlock (*Monograptus flexilis* zone) to top Přídolí (*Monograptus uniformis* zone) and into the Lower Devonian (*Monograptus hercynicus* zone). These latter sections through the Silurian-Devonian boundary are only found 2 km north of Villaputzu on a hill called Baccu Scottis. Some thirty metres of limestone separates the upper black graptolitic shales from those of Lower Ludlow age. At the top of the limestones it may be possible to find the bulbous root structures (?floats) of the crinoid *Scyphocrinus* that is considered to be planktonic in habit. Note that Sardinian summers can be very hot and spring or autumn are more suitable for graptolite hunting in these hills. For a sketch map see Jaeger, H., 1977 'The Silurian-Devonian boundary in Thuringia and Sardinia', *in* Martinsson, A. (ed.) *The Silurian-Devonian Boundary*. IUGS Series A, No. 5, pp. 117–125. Also guidance may be obtained from the Department of Earth Sciences in the University at Cagliari, via Trentino 51.

CANADA

One of the most famous graptolite localities, made so by James Hall (see biographical sketch on p. 142) is also one of the most accessible at Lévis near Quebec City. Advice can be obtained from the Department de Géologie, Université Laval. Numerous localities occur in the region; and also in the Utica Shale along the St Lawrence River, southwest of Quebec City.

USA

ALABAMA AND TENNESSEE. Middle Ordovician (Athens Shale) graptolites are well preserved, abundant and collectable at Pratt's Ferry and Pratt's Syncline, 10 miles NE of Centerville on route 82; and at a quarry 4 miles W of Calera. Both localities are just N of route 25, Pratt's Ferry being close to route 27 on the west side. Graptolites belonging to seventeen taxa are known from these sites and include *Pterograptus*, *Azygograptus*, *Didymograptus*, *Thamnograptus* and *Reteograptus*. Further details can be found in Finney, S. C., 1980, *Journal of Paleontology* 54, 1184–1208, and in other references quoted in that publication.

ARKANSAS AND OKLAHOMA. Good dendroid graptolites and Caradoc (approximately *gracilis* Zone) graptolites can be collected at several sites of Normanskill Shale: NE from Caddo Gap on the road between Plata and Ophir; black shale interbedded with blue limestone at Crystal Springs, Montgomery County; 12–13 miles NW of Hot Springs; just below the dam at Caddo Gap. Llandovery, Silurian graptolites have been obtained on the Little Missouri River at the S base of Blaylock Mountain, Caddo Gap quadrangle. The Ordovician Viola Limestone Fm of Oklahoma is several hundred feet thick in southern Oklahoma and is famous for its graptolites at several localities. Further localities and faunal details are listed in Ruedemann, R., 1947, *Geological Society of America*, Memoir 19, 1–652; and Decker, C., 1936, *American Association of Petroleum Geologists* 20, 301–311.

IDAHO AND NEVADA. Arenig to Ashgill, Ordovician graptolites, well preserved and of varied kind, (*Dicellograptus, Climacograptus*, retiolites) can be collected near Trail Creek, Idaho and Chicken Creek, Nevada. The Trail Creek localities are located on the ridge crest half a mile NW of the summit of the Trail Creek – Mackay Road. They occur in black shales of the Phi Kappa Formation. The forms listed above are Caradoc forms but nearby localities yield specimens of other Ordovician series. Details can be found in Churkin, M., 1963, *American Association of Petroleium Geologists* 47, 1611–1623; and in Carter, C., 1972, *Journal of Paleontology* 46, 43–49.

MAINE. End Ordovician (*persculptus* Zone) graptolites can be collected near Caribou in Co. Maine. The locality is in a road cut on Route 161, some 4 miles (6.5 km) NW of Caribou. The graptolites occur in a 25 cm thick hard, silty shale band intercalated between 'ribbon rock' limestone below and a conglomerate above. The stratigraphic formation is the Carys Mills Formation. Details can be found in Rickards, R. B. & Riva, J., 1981, *Geological Journal* 16, 219–235.

MISSOURI. Late Ordovician biserial graptolites (*Orthograptus* and *Climacograptus*) can be collected at localities near Barnhart 25 miles SW of St Louis, and Castlewood 20 miles WSW of St Louis: the respective highways are routes I–55 and I–44. The Castlewood localities are close to the Meremec River; the Barnhart localities nearer Barnhart than the Meremec River. The specimens show details of virgella and nema, and are found in shale interbedded between the limestone of the Kimmswick Formation and the sandstone of the Bushberg Formation. The shale with graptolites is referred to the Maquoketa Formation. Further information can be obtained from Berry, W. B. .N. & Marshall, F. C., 1971, *Journal of Paleontology* 45, 253–257.

MONTANA AND WYOMING. A few localities have yielded varied Cambrian and Ordovician (Deepkill) dendroid graptolites, including *Dictyonema, Dendrograptus, Acanthograptus, Airograptus, Cactograptus* and *Inocaulis*. These localities are: N side of Shoshone Canyon, just above the bridge, west of Cody, Wyoming; quarter of a mile west of Stillwater River, Montana (in the Bighorn Limestone). Details, and some illustrations of these dendroids, can be found in Ruedemann, R., 1947, *Geological Society of America*, Memoir 19, 1–652.

OHIO AND INDIANA. Graptolites are not common in these States but have been reported from a few localities (see references at the end of this section), the most dramatic in terms of fine preservation being on the S side of Stonelick Lake, near the sewage plants in Stonelick Lake State Park, Clermont County, some 30 miles ENE of Cincinnati. A thin sequence of shale of Upper Ordovician age has the graptolites most concentrated in somewhat dolomitised calcareous bands and mudstone. Details can be found in Mitchell, C. E. & Bergström, S. M., 1977, *Bollettino della Società Italiana* 16, 257–270; and information about other localities in Ohio are in Ruedemann, R., 1947, *Geological Society of America*, Memoir 19, 1–652 and Berry, W. B. N., 1966, *Journal of Paleontology* 40, 1392–1394.

TEXAS. In the Marathon region of west Texas are several sections where numerous graptolites can be collected from varied rock types – shales, cherts, limestones and sandstones. They are of Ordovician age (Tremadoc to Ashgill Series) and very diverse generically. Some of the best sections are as follows: 4 miles SW of Marathon (bearing S65°E); 9.5 miles S35°W of Marathon, on east side of gap between ridges of Marathon limestone; at a cairn 6.6 miles S55°W of Marathon, on Roberts Ranch road where the road turns SW past Alsate Creek. More details can be found in Berry, W. B. N., 1960, University of Texas publication number 6005, 1–179.

VIRGINIA. *Gracilis* Zone, Caradoc, Ordovician faunas are well known and diverse in this State. In SW Virginia the Catawba Valley section has yielded a rich assemblage including *Nemagraptus, Didymograptus, Climacograptus, Diplograptus* and *Glossograptus*. Other localities are at Wytheville (reservoir excavations) and 5 miles E of Saltville in Porterfield Quarry. Further information occurs in Powell, S. L., 1915, *Journal of Geology* 23, 272–281; and Ruedemann, R., 1947, *Geological Society of America*, Memoir 19, 1–652.

YUKON. Although for most people inaccessible, good Ordovician to Devonian graptolite localities are available in the Road River Formation outcrops near the Yukon River on the western edge of the Ogilvie Mountains (Alaska), some 75 miles NW of Dawson, between Dawson and Woodchopper. The Road River Fm is essentially graptolitic shale and chert and is underlain by Cambro-Ordovician limestones and overlain by Upper Devonian clastics. Silurian/Devonian boundary collections of monograptids have been made at a site 4 miles due N on the confluence of the Salmontrout and Porcupine rivers (in Alaska, 170 miles to the N of the Yukon River region mentioned above). Full details can be found in Churkin, M. & Brabb, E. E., 1968, *International Symposium on the Devonian System* 2, 227–258, Calgary.

AUSTRALIA

This country has some of the finest collecting localities in the world, the Bendigo region, north of Melbourne perhaps being the most famous. There are certainly in excess of 1000 known localities in the Bendigo district. Guidance would be needed from the National Museum of Victoria (in Melbourne), the Mines Department or the University of Melbourne, each of which holds comparative collections. The regions west of Sydney, around Orange, but also near Parkes, Forbes, Yass and Goulburn, have numerous graptolite localities of varied quality and ease of collecting, but in these cases it is necessary to get an 'in' to the literature before going into the field. The Universities of Sydney or Wollongong could give help.

Appendix 3

HOW DO WE CLASSIFY THEM?

The classification of graptolites has received considerable attention in recent years, the most notable being by Rigby (1986, Special Publications of the Geological Society, 20, 1–13), Fortey and Cooper (1986, *Palaeontology*, 29, 631–654) and Mitchell (1987, *Palaeontology*, 30, 353–405). The first paper discusses the history of classification, paying special attention to the development of the Chinese, Russian and Western European systems of classification. The second and third papers, largely followed in this appendix, introduce the latest concepts of phylogenetic classifications of graptolites. In the classification over the page Figure references are given to the photographs, where examples of the group are discussed or illustrated. The *circled* numbers are the same as those used in Appendix 4.

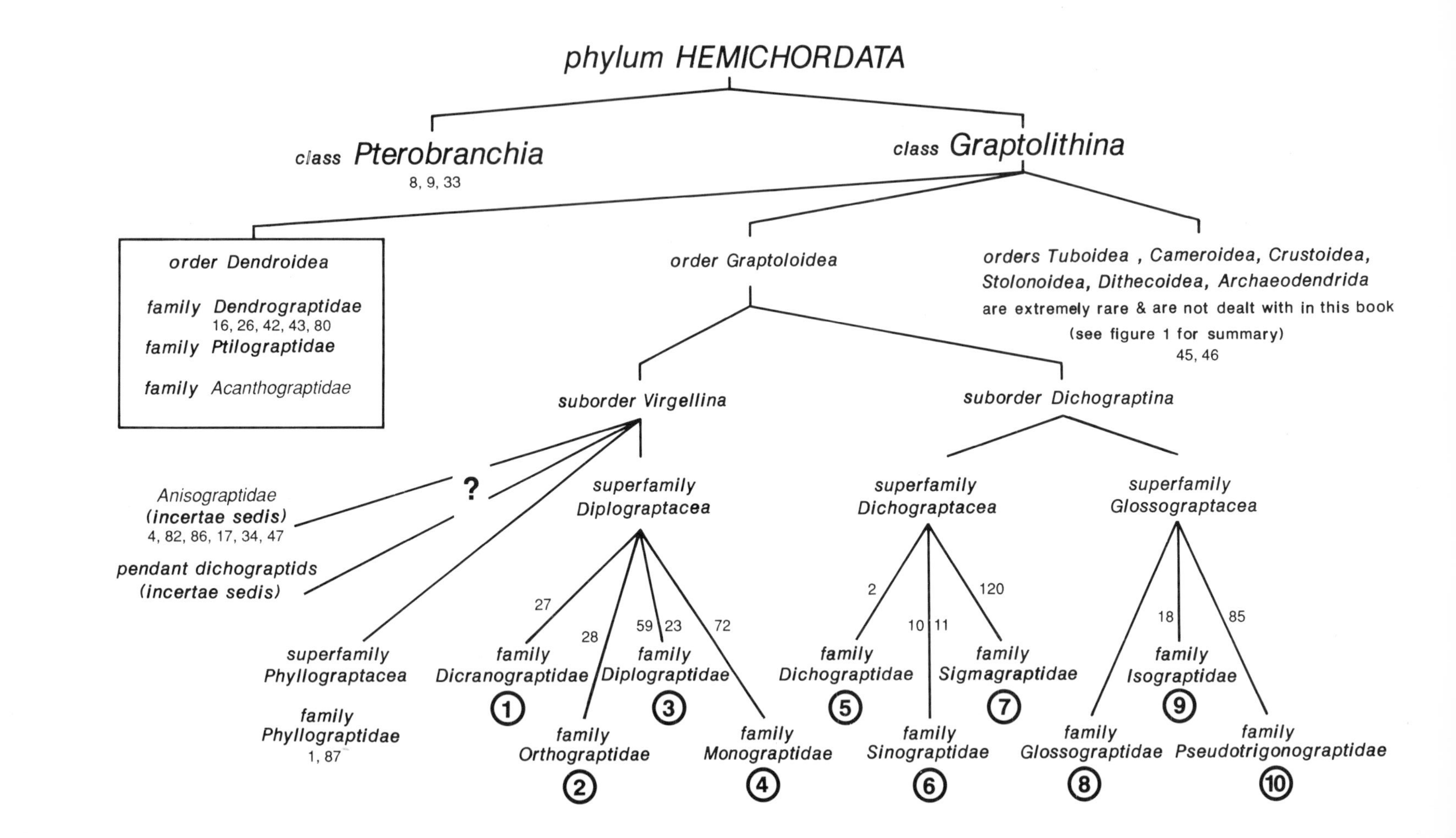

phylum HEMICHORDATA
class Pterobranchia
8, 9, 33
class Graptolithina
order Dendroidea
family Dendrograptidae
16, 26, 42, 43, 80
family Ptilograptidae
family Acanthograptidae
order Graptoloidea
orders Tuboidea , Cameroidea, Crustoidea, Stolonoidea, Dithecoidea, Archaeodendrida
are extremely rare & are not dealt with in this book
(see figure 1 for summary)
45, 46
suborder Virgellina
suborder Dichograptina
?
Anisograptidae
(incertae sedis)
4, 82, 86, 17, 34, 47
pendant dichograptids
(incertae sedis)
superfamily Phyllograptacea
family Phyllograptidae
1, 87
superfamily Diplograptacea
superfamily Dichograptacea
superfamily Glossograptacea
27
28
59
23
72
2
10
11
120
18
85
family Dicranograptidae
1
family Orthograptidae
2
family Diplograptidae
3
family Monograptidae
4
family Dichograptidae
5
family Sinograptidae
6
family Sigmagraptidae
7
family Glossograptidae
8
family Isograptidae
9
family Pseudotrigonograptidae
10

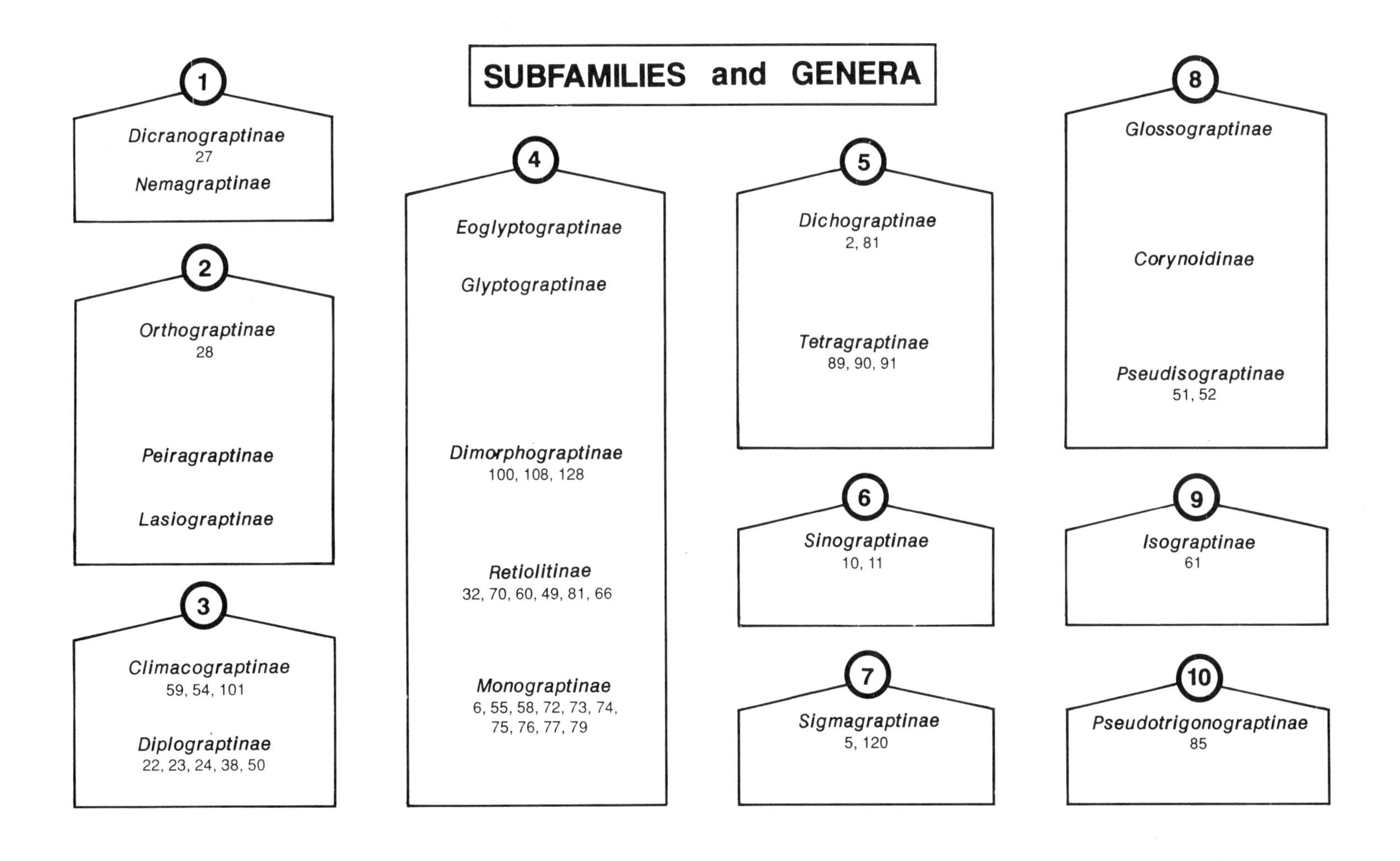
SUBFAMILIES and GENERA
1
Dicranograptinae
27
Nemagraptinae
2
Orthograptinae
28
Peiragraptinae
Lasiograptinae
3
Climacograptinae
59, 54, 101
Diplograptinae
22, 23, 24, 38, 50
4
Eoglyptograptinae
Glyptograptinae
Dimorphograptinae
100, 108, 128
Retiolitinae
32, 70, 60, 49, 81, 66
Monograptinae
6, 55, 58, 72, 73, 74,
75, 76, 77, 79
5
Dichograptinae
2, 81
Tetragraptinae
89, 90, 91
6
Sinograptinae
10, 11
7
Sigmagraptinae
5, 120
8
Glossograptinae
Corynoidinae
Pseudisograptinae
51, 52
9
Isograptinae
61
10
Pseudotrigonograptinae
85

APPENDIX 4

HOW DID THEY EVOLVE?

The evolution of graptolites is a subject more or less confined at present to the study of graptoloids: little is known of the dendroids and other orders (Fig. 1) whether of relationships within the orders or of connections between them, despite a wealth of knowledge, often, of their detailed morphologies. By contrast the graptoloids afford one of the best evolutionary accounts in the fossil record, not least because of their fine stratigraphic documentation. The most recent major works dealing with the evaluation of graptolites are those referred to in Appendix 3, on classification, for they take a fundamentally phyllogenetic approach; and those listed in Appendix 5, referring to Chapter 12. This table is a state-of-the-art attempt; an attempt to make a best fit to the recent concensus views. By following some of the references given the reader can trace further the ideas on evolution of actual lineages, that is at a lower taxonomic level than given in the chart. On the chart Figure references are given to photographs in this book which illustrate that particular group. Excluded from the chart are the pendent dichograptids (see Appendix 3) and the retiolids. The evolutionary framework depicted is largely after Rickards *et al.* 1977. *Bull. Brit. Mus.* (*Nat. Hist.* 28, 1–120), Fortey and Cooper (1986) and Mitchell (1987). *Circled* numbers used on the chart are the same as those on the classification chart of Appendix 3.

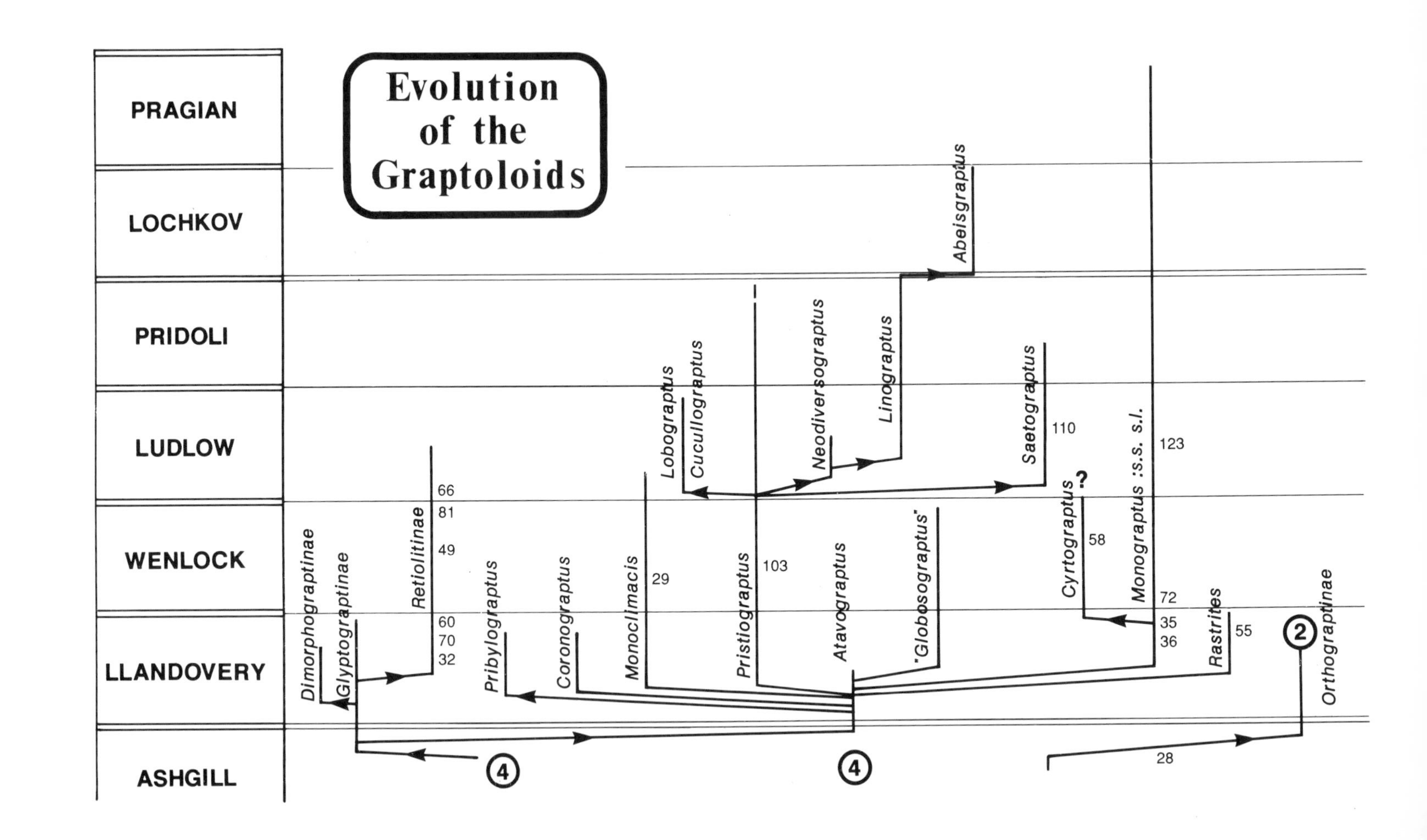

Evolution of the Graptoloids
PRAGIAN
LOCHKOV
PRIDOLI
LUDLOW
WENLOCK
LLANDOVERY
ASHGILL
Dimorphograptinae
Glyptograptinae
Retiolitinae
66
81
49
60
70
32
Pribylograptus
Coronograptus
Monoclimacis
29
Lobograptus
Cucullograptus
Pristiograptus
103
Neodiversograptus
Linograptus
Abeisgraptus
Atavograptus
'Globosograptus'
Saetograptus
110
Cyrtograptus
?
58
Monograptus :s.s. s.l.
123
72
35
36
Rastrites
55
Orthograptinae
2
4
4
28

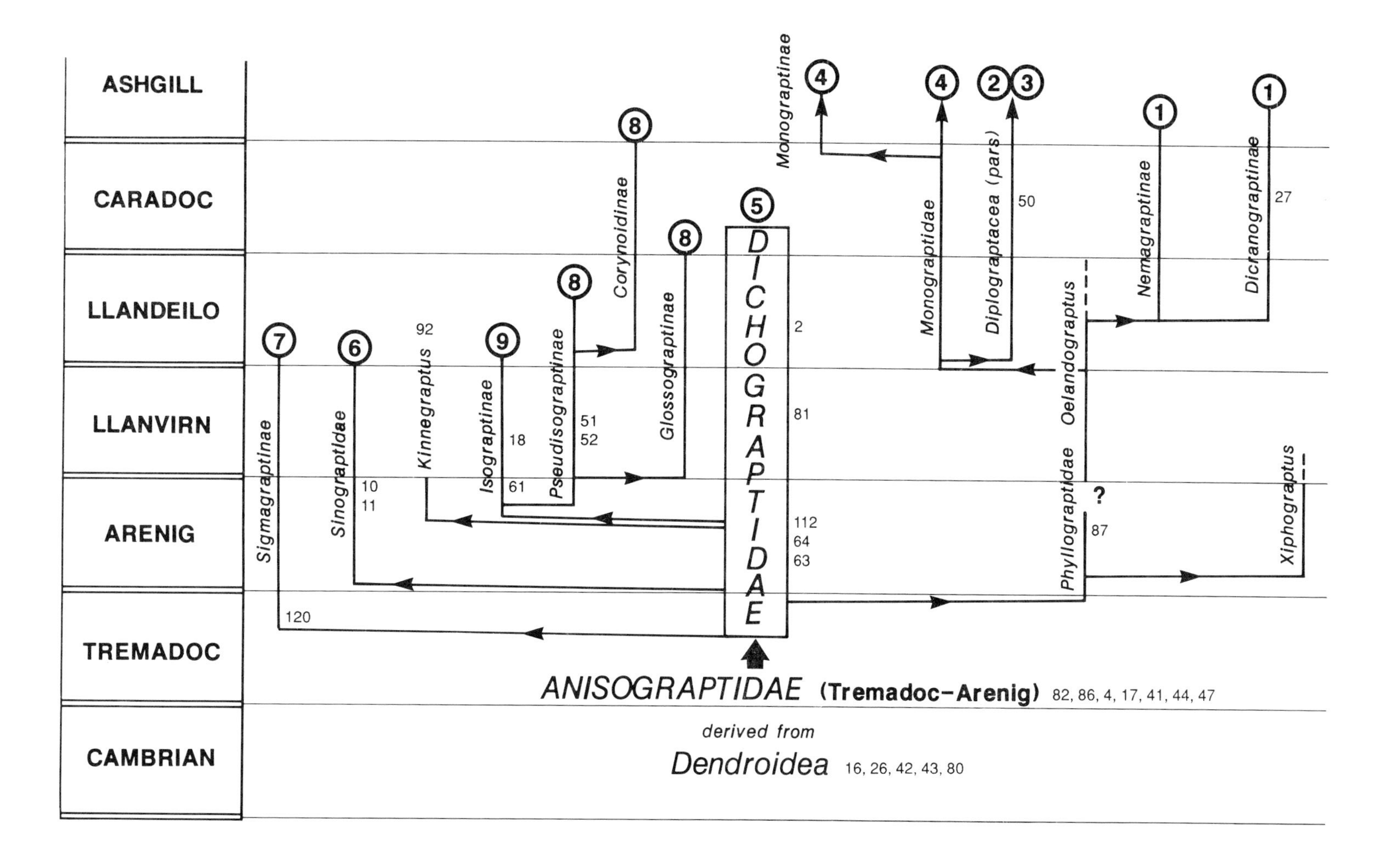
ASHGILL
CARADOC
LLANDEILO
LLANVIRN
ARENIG
TREMADOC
CAMBRIAN
Sigmagraptinae
120
Sinograptidae
10
11
Kinnegraptus
92
Isograptinae
18
61
Pseudisograptinae
51
52
Corynoidinae
Glossograptinae
DICHOGRAPTIDAE
2
81
112
64
63
Monograptinae
Monograptidae
Diplograptacea (pars)
50
Oelandograptus
Phyllograptidae
?
87
Nemagraptinae
Dicranograptinae
27
Xiphograptus
ANISOGRAPTIDAE (Tremadoc–Arenig) 82, 86, 4, 17, 41, 44, 47
derived from
Dendroidea 16, 26, 42, 43, 80

APPENDIX 5

WHERE CAN I GO FOR HELP?

For anyone who wants to follow up an interest in graptolites especially by collecting specimens and then trying to identify them, there are basically three different approaches. There is the use of established fossil collections, the palaeontological literature or through the organisations of palaeontology and geology. Ideally a combination of all three should be used. This appendix is intended as a brief introduction and practical guide to these three areas. For a more detailed and international guide that lists palaeontological organisations, journals and museums on a worldwide basis see Nudds and Palmer, 1989 (reference Appendix 5). In this appendix we are aiming at helping the amateur gain an entrée into what might seem to be a rather exclusive and specialised professional world. Do not be put off; most palaeontologists want to encourage the amateur and subscribe to the more literal and old fashioned meaning of 'amateur' rather than its modern slightly derogatory connotation.

Firstly collections: there is no doubt that for the inexperienced collector to identify graptolites or any fossils successfully, it is necessary to make comparisons with reliably identified specimens as well as using published descriptions and illustrations. Unfortunately there are very few comprehensive illustrated keys available for the identification of fossils compared with other areas of natural history such as ornithology or lepidoptery. Consequently it is helpful to be able to use open access or public fossil collections and this means having access to museum collections.

The first step would be to try your local museum. There is a surprising number (more than 150) around the country with geological collections of some kind, although only a small proportion have a geologist on the staff, unfortunately. Even if they do not, ask the curator anyway; they are often very knowledgeable and helpful and will redirect you to the nearest appropriate source if they cannot help you themselves.

If you are not sure where the local museum is, ask in the public library and consult **The Geologist's Directory** (4th edition E. McInairnie, 1988, published by the Institution of Geologists, i–x 219pp., ISBN 0 950 6906 51, and available from Burlington House, Piccadilly, London W1V 9HQ. This is an extremely useful, annually updated source of information that lists geological museums, libraries, societies, journals, university departments etc. with contact addresses etc. The information on museums gives approximate size of collections and whether there is a geologist on the staff. See also Cleevely's **World Palaeontological Collections**

(1983, British Museum, Natural History Publications) for an international list.

Having ascertained that the local museum does have a fossil collection with graptolites in it, there are some questions that need to be asked. Are they identified? When were they identified and by whom? If it is an old collection (i.e. more than 60 years old) then the generic names will be out of date and not very helpful although the species names may be accurate enough. If they were recently identified then it is possible that the person who made the identifications is still around and may be able to help. Not only that but they may have some of the appropriate literature that will help you identify your own fossils.

Where do they come from and are they accurately located? If they are locally derived (check the geological map to see if there is any likelihood of there being local graptolites) then they may guide you to a place that would be worth recollecting (see **Chapter 11** on collecting graptolites). If the local museum does not have graptolites in its fossil collections then offer to collect some for them. The best place for a good private collection is in a museum where they can be properly curated and conserved for future generations.

Since there is not yet a national directory of fossil collections in local museums, the following is a list of larger museums that do have graptolite collections.

NATIONAL MUSEUMS

The national museums all have extensive fossil collections but the representation of graptolites within them varies enormously. However, these museums do all have palaeontologists and part of their work is to assist members of the public with their enquiries.

British Museum (Natural History) and **Geological Museum**, Cromwell Road, London SW7 5BD (Tel. 01 938 9123). Extensive collections, British and foreign, one of the best collections of graptolites in the country.

National Museum of Wales, Cathays Park, Cardiff CF1 3NP (Tel. 0222 397951). Extensive collections especially from Wales and the Welsh Borders.

National Museum of Scotland, Chambers Street, Edinburgh EH1 1JF (Tel. 01 225 7534). Graptolite collections from the Southern Uplands of Scotland.

Ulster Museum, Botanic Gardens, Belfast, Northern Ireland, BT9 5AB (Tel. 0232 668251). Graptolite collections from the Ards Peninsula and particularly Coalpit Bay, Donaghadee, and from other parts of Ireland. These were all identified by Rickards in modern times so will have on them more or less up to date names.

Just as important and probably even more important from a national point of view are the major collections of fossils held by the **British Geological Survey**. These are the most extensive and representative collections of the palaeontology of the British Isles and are housed in Keyworth and Edinburgh. Needless to say, the graptolite collections are very good and result from over 150 years of field collecting by innumerable geologists. The public are encouraged to make use of the collections by appointment. The English and Welsh specimens are at the British Geological Survey, Keyworth, Nottinghamshire NG12 5GG, (Tel. Plumtree [06077] 6111) and Scottish specimens at the British Geological Survey (Scottish Office) Murchison House, West Mains Road, Edinburgh EH9 3LA (Tel. 231–667 1000).

UNIVERSITY MUSEUMS

There are some University museums that have outstanding collections of graptolites because of their association with particular graptolite researchers. The eminent amongst these are the:

Sedgwick Museum, University of Cambridge, Downing Street, Cambridge CB2 3EQ (Tel. 0223 333437). Extensive world-wide collections, but particularly comprehensive British collections made by G. L. Elles and E. M. R. Wood, O. M. B. Bulman, R. B. Rickards and many others.

University of Birmingham, Department of Geological Sciences, PO Box 363, Birmingham B15 2TT (Tel. 021 472 1301, access by appointment). Extensive British collections made by C. Lapworth, G. L. Elles and E. M. R. Wood, and I. Strachan.

University College of Wales, Department of Geology, Aberystwyth, Dyfed SY23 3QB (Tel. 0970 3711, access by appointment). Predominantly Welsh collections, especially from the Rheiddol Gorge.

Hunterian Museum, University of Glasgow, Glasgow G12 8QQ (Tel. 041 330 4221). Mainly Scottish material.

Many other university departments of Geology have fossil collections and some have small museums. Often these collections will contain graptolites but since they are primarily used for teaching and research access is by appointment (See **The Geologist's Directory** for details).

PALAEONTOLOGICAL PUBLICATIONS

Unfortunately there is no straightforward illustrated key for the identification of graptolites or indeed for most other fossil animal groups in the same way that there are floral keys. There are a few well illustrated palaeontological reference books that include graptolites e.g.:

The Atlas of Invertebrate Palaeontology, 1985, Murray, J. (ed.) The Palaeontological Association, Longmans. This is a recent text, well illustrated with photos and brief descriptions of some genera throughout the whole of the Class Graptolithina and has a general introduction.

British Palaeozoic Fossils, (numerous editions) published by the Trustees of the **British Museum (Natural History)**. Bookshops usually have only the latest editions. Drawings of the more common fossils, including graptolites, arranged stratigraphically. Drawings somewhat approximate.

There are a number of good modern general palaeontological textbooks that have illustrated chapters on all aspects of graptolites:

The Elements of Palaeontology, 1988, Black, R. M. Cambridge University Press xii, 404pp. An 'A' level and 1st year university text with accurate information on graptolites.

Invertebrate Palaeontology and Evolution, 1979, 1986, (2nd edition) Clarkson, E. N. K. Allen & Unwin. One of the most widely used undergraduate text in Britain with an up to date chapter on graptolites.

Fossil Invertebrates, 1987, Cheetham, A. H. and Rowell, A. J. (eds). Blackwell Scientific Publications, Palo Alto and Oxford, xi, 713pp. Recent multi-author, advanced undergraduate text with chapter on graptolites by American graptolite specialist W. B. N. Berry.

MORE SPECIFIC AND ADVANCED REFERENCES ON GRAPTOLITES

A major data source for invertebrate palaeontology is the multi-volume **Treatise on Invertebrate Palaeontology**. This is an important and continually updated American series that attempts to illustrate and classify all valid fossil taxa of genus and higher rank. The late O. M. B. Bulman compiled the volume dealing with the Class Graptolithina. This is a detailed technical introduction to all aspects of graptolite studies with illustrations and descriptions of many genera. It is a graduate level text and an essential reference for detailed investigation.

Graptolithina, 1970, Bulman, O. M. B. *In* Teichert, C. (ed.) **Treatise on Invertebrate Palaeontology**, Part V (2nd edition, revised). The Geological Society of America and the University of Kansas, Boulder, Colorado and Lawrence, Kansas i–xxxii, 1–163.

For identification at the species level the major compilatory works are the monographic publications of the British Palaeontographical Society, one of the oldest continuing series of this kind in the world (from 1848). The British graptoloid graptolite fauna was described by Gertrude Elles and Ethel Wood, both students of Charles Lapworth, between 1901 and 1918 and the dendroid fauna by O. M. B. Bulman between 1929 and 1967. Since then parts of the fauna have been revised and expanded by Bulman (1944), Strachan (1971), Rickards (1970), Hutt (1974, 1975) and Hughes (1989).

Monograph of British Graptolites, Parts I–XI, 1901–1918, Elles, G. L. and Wood, E. M. R. Palaeontographical Society [Monograph], clxxi, 539pp.

British Dendroid Graptolites, Parts I–IV, 1927–1967, Bulman, O. M. B. Palaeontographical Society [Monograph], lxiv, 97pp.

Caradoc (Balclatchie) Graptolites from limestones in Laggan Burn, Ayrshire, 1944, Bulman O. M. B. Palaeontographical Society [Monograph], 78pp.

The Llandovery (Silurian) Graptolites of the Howgill Fells, Northern England, 1970, Rickards, R. B. Palaeontographical Society [Monograph], 108pp.
A Synoptic Supplement to 'A Monograph of British Graptolites by Miss G. L. Elles and Miss E. M. R. Wood', 1971, Strachan, I, Palaeontographical Society [Monograph], 130pp.
The Llandovery Graptolites of the English Lake District, parts I & II, 1974, Hutt, J. E., Palaeontographical Society [Monograph], 137pp.
Llandeilo and Caradoc Graptolites of the Builth and Shelve Inliers, 1989, Hughes, R. A., Palaeontographical Society [Monograph], 89pp.

REGIONAL AND LOCAL GEOLOGICAL GUIDES

The British Geological Survey is one of the oldest established surveys in the world and has been publishing geological maps and descriptions of the regional geology for nearly 150 years. Some of these are very well written and readily accessible to the amateur palaeontologist. This is especially true of the regional guide series which separately describes the different geological regions of the British Isles. In them the stratigraphy and faunas of the successive periods of geological time are given with specific local examples. Consequently it is possible to extract details of particular graptolitic horizons and localities that can be visited and collected (*with permission*, see **Chapter 11**). The relevant guides to graptolite bearing Lower Palaeozoic strata are those for Wales and the Welsh Borders, Southern Scotland and Northern England.

More detailed accounts of local geology are contained in the technical memoirs that accompany the one inch to the mile and now 1:50,000 geological maps. Individual fossil localities and sections indicated on the maps are described in the memoirs with their faunas. Together these publications are an invaluable source of information. To find out which maps and memoirs deal with the area you are interested in, consult the appropriate regional guide.

The various modern general palaeontological texts and HMSO publications mentioned above can be obtained or at least ordered from any good book shop. Certain 'classic' areas are the subject of detailed maps and popular handbooks (e.g. Cross Fell). Also, it may be possible to persuade a local library to obtain copies that can be borrowed. Again using the national library service it is possible to borrow copies of the more specialised texts from the British Library. Of course they have to be used in your branch library but they are obtainable. Some large town or city libraries may have a reasonable selection of palaeontological texts and even some geological journals, especially where there is a local third level educational institution teaching geology. However, the main library sources of palaeontological publications are the libraries associated with those same geology teaching departments, museums and geological societies. Access to most of these is restricted and by appointment or membership only but *bona fide* enquiries are rarely turned away. For a list of some of the libraries throughout the British Isles with geological holdings see The Geologist's Directory.

Beyond these few selected works, the vast bulk of the specialised literature dealing with graptolite faunas from other parts of the world, all the different aspects of their palaeobiology, ecology and stratigraphy is scattered through a vast array of books and journals published over the last 150 years and more, and originating from many different countries. It has been estimated that there are at least 6000 separate publications (mainly papers in journals) dealing specifically with graptolites. Almost all of these are specialist texts that can be difficult to obtain for consultation, although references to them are given in the Palaeontographical Society monographs listed above.

HOW TO GAIN ACCESS TO THE SPECIALIST GRAPTOLITE LITERATURE

Firstly, it is necessary to gain entrance to a good geological library but having done that how does one set about finding what is wanted from amongst the rows of books and journals that will certainly be catalogued but not indexed?

Perhaps the major difficulty is knowing exactly what it is that you want. If it is just a matter of trying to get some information on a clearly defined topic, such as the graptolite faunas of a specific locality e.g. Abereiddi Bay in Wales or stratigraphic horizon (e.g. the *Monograptus ludensis* zone at the top of the Wenlock) then there are straightforward bibliographic procedures that can be followed. If however information is required on a more general or less well defined topic, for instance, 'what is known about how graptolites fed or reproduced?' then it could be difficult to find. This is because the subject area and its Key words *feeding* and *reproduction* would not be identifiable in most geological bibliographies. It is necessary to establish a heirarchy of appropriate Key words and work from the narrowest and most exact subject areas to the broader ones until the right references are found.

There is no doubt that the quickest and most efficient way to find what you want is to ask a sympathetic and trained librarian for help. Failing this, the standard methodical bibliographic search procedures have to be followed. For palaeontology the most extensive English language bibliography is the American **Bibliography and Index of Geology**, which draws its data from nearly 3000 journals, books etc. and is published in monthly parts. It will only be available in specialised geological libraries but it is also computerised and available 'on-line' as **GeoRef**. However, in this form it is quite expensive to use.

THE ORGANISATIONS OF PALAEONTOLOGY AND GEOLOGY

The main British based palaeontological organisation is the **Palaeontological Association** (for detailed information on membership write to Dr H. Armstrong, Geology Department, The University, Newcastle upon Tyne, NE1 7RU) which promotes research in all aspects of palaeontology, and has a worldwide membership. It publishes a quarterly international journal (*Palaeontology)* (and also various monograph works and fossil guides) and Newsletters that are issued free to members. The association holds a peripatetic Annual Conference each December at one of the Universities (Dublin 1991), organises review seminars, lecture meetings and field excursions throughout the year. Non-members are welcome to most of these activities and the Association actively seeks to encourage amateurs.

There is also a whole range of more general geological societies on the national, regional and local level which help promote palaeontological interests as part of their activities. Like the Palaeontological Association many of them hold regular meetings and field excursions, publish journals and some have libraries and small museums.

NATIONAL ORGANISATIONS

The Geologists' Association, Burlington House, Piccadilly, London W1V 0JU (Tel. 01 784-2356).

The Geological Society, Burlington House, Piccadilly, London W1V 0JU (Tel. 01 734 2356).

Royal Irish Academy, 19, Dawson Street, Dublin 2, Ireland (Tel. Dublin 762570).

Royal Society of Edinburgh, 22, George Street, Edinburgh EH2 2PQ, Scotland (Tel. 031 225 6057).

REGIONAL ORGANISATIONS

Belfast Geologists' Association, 6, Gibson Park Avenue, Gregagh, Belfast 6, Northern Ireland.

Edinburgh Geological Society, British Geological Survey, West Mains Road, Edinburgh EH9 3LA.

Geological Society of Glasgow, Department of Geology, The University, Glasgow G12 8QQ.

The Geologists' Association, Burlington House, Piccadilly, London W1V 0JU (has a number of local groups throughout England and Wales, contact secretary for further information).

Irish Geological Association, c/o Geology Department, University College, Belfield, Dublin 4, Ireland.

Liverpool Geological Society, Geology Section, Liverpool Polytechnic, C.F. Mott Campus, Liverpool Road, Prescot L34 1NP.

Royal Geological Society of Cornwall, Geological Museum, Alberton Street, Penzance, Cornwall TR18 2QR.

The Ussher Society, c/o British Geological Survey, St Just, 30 Pennsylvania Road, Exeter, Devon EX4 6DT.

Yorkshire Geological Society, Department of Geology, University of Sheffield, Beaumont Building, Brook Hill, Sheffield S13 7HF.

This is by no means an inclusive list, there are many other local geographical and natural history societies throughout the British Isles, for a more comprehensive list see **The Geologist's Directory**.

Appendix 6

WHERE CAN I READ MORE ON THE SUBJECT?

We have given below all the references quoted in the text, arranged chapter by chapter, but we have in addition listed useful papers not specifically quoted. When coupled with Appendix 6 on where to go for help, both the amateur and professional geologist should be able to make progress, as, hopefully, will teachers at all levels. Clearly there is some duplication of references from chapter to chapter, for many subject areas overlap and many definitive papers discuss several subject areas.

CHAPTER ONE

Bates, D. E. B. and Kirk, N. H., 1987. The role of extrathecal tissue in the construction and functioning of some Ordovician and Silurian retiolitid graptoloids. *Bulletin of the Geological Society of Denmark*, 35, 85–102.

Bohlin, B., 1950. The affinities of the graptolites. *Bulletin of the Geological Institution of the University of Uppsala*, 34, 107–113.

Bulman, O. M. B., 1970. Graptolithina with sections on Enteropneusta and Pterobranchia. *In* Teichert, C. (ed.), *Treatise in Invertebrate Palaeontology, Part V (2nd Edition).* Geological Society of America and University of Kansas Press. i–xxxii, 1–163.

Crowther, R. R. and Rickards, R. B., 1977. Cortical Bandages and the graptolite zooid. *Geologica and Palaeontologica*, 11, 9–46.

Dilly, P. N., 1986. Modern Pterobranchs, Observations on their behaviour and tube building. *In* Hughes, C. P., Rickards, R. B., Chapman, A. J., (eds), Palaeoecology and Biostratigraphy of Graptolites. 261–269. *Geological Society Special Publication No. 20.*

Elles, G. L. and Wood, E. M. R., 1901–18. British Graptolites. *Palaeontographical Society Monographs*, 1–clxxi, 1–359.

Fortey, R. A. and Owens, R. M., 1987. The Arenig Series in South Wales: Stratigraphy and Palaeontology. *Bulletin of the British Museum (Natural History) Geology*, 41, 69–307.

Holland, C. H., Rickards, R. B. and Warren, P. T., 1969. The Wenlock graptolites of the Ludlow District, Shropshire, and their stratigraphical significance. *Palaeontology*, 12, 663–683.

Hughes, R. A., 1989. Llandeilo and Caradoc Graptolites of the Builth and Shelve Inliers. *Palaeontographical Society Monographs*, 141, 1–89.

Hutt, J. E., 1974. The Llandovery Graptolites of the English Lake District. Part I *Palaeontographical Society Monographs*, 128, 1–56.

Hutt, J. E., 1975. The Llandovery Graptolites of the English Lake District. Part 2 *Palaeontographical Society Monographs*, 129, 57–137.

Kozlowski, R., 1948. Les graptolithes et quelques nouveaux groupes d'animaux du Tremadoc de la Pologne. *Palaeontologica polonica* 3, 1–235.
Rickards, R. B., 1965. New Silurian graptolites from the Howgill Fells (Northern England) *Palaeontology*, 8, 247–271.
Rickards, R. B. and Archer, J. B., 1969. The Lower Palaeozoic Rocks near Tomgraney, Co. Clare. *Scientific Proceedings of the Royal Dublin Society*, A, 3, 219–230.
Rickards, R. B., Chapman, A. J. and Temple, J. T. *Rhabdopleura hollandi*, a new pterobranch hemichordate from the Silurian of the Llandovery district, Powys, Wales. *Proceedings of the Geologists Association*, 95, 23–28.
Rickards, R. B. and Dumican, L. W., 1984. The fibrillar component of the graptolite periderm. *The Irish Journal of Earth Sciences*, 6, 175–203.
Williams, S. H., 1982. Upper Ordovician graptolites from the top Lower Hartfell Shale Formation (*D. clingani* and *P. linearis* zones) near Moffat, southern Scotland. *Transactions of the Royal Society of Edinburgh: Earth Sciences*, 72, 229–255.

CHAPTER TWO

Dumican, L. W. and Rickards, R. B., 1985. Optimum preparation, preservation and processing techniques for graptolite electron microscopy. *Palaeontology*, 28, 757–766.
Kozlowski, R., 1948. Les graptolithes et quelques nouveaux groupes d'animaux du Tremadoc de la Pologne. *Palaeontologica polonica*, 3, 1–235.
Rickards, R. B. and Stait, B. A., 1984. *Psigraptus*, its classification, evolution and zooid. *Alcheringa*, 8, 101–111.
Williams, S. H., 1981. Form and mode of life of *Dicellograptus* (Graptolithina). *Geological Magazine*, 118, 401–408.
Williams, S. H. and Rickards, R. B., 1984. Palaeoecology of graptolitic black shales. *In* Bruton, D. I., (ed.), *Aspects of the Ordovician System*, 159–166. Palaeontological contributions from the University of Oslo, no. 295, Universitetsforlaget.

CHAPTER THREE

Armstrong, W. G., Dilly, P. N. and Urbanek, A., 1984. Collagen in the pterobranch coenecium and the problem of graptolite affinities. *Lethaia*, 17, 145–152.
Rickards, R. B. and Dumican, L. W., 1984. The Fibrillar component of the graptolite periderm. *Irish Journal of Earth Sciences*, 6, 175–203.
Rickards, R. B. and Dumican, L. W., 1985. Graptolite ultrastructure: evolution of descriptive and conceptual terminology. *Geological Magazine*, 122, 125–137.

CHAPTER FOUR

Cooper, R. A. and Fortey, R. A., 1982. The Ordovician graptolites of Spitsbergen. *Bulletin of the British Museum (Natural History)* Series 36, 157–302.
Koren', T. N. and Rickards, R. B., 1979. Extinction of the graptolites. *In* Harris, A. L., Holland, C. H. and Leake, B. E., (eds), The Caledonides of the British Isles – reviewed. 457–466. *Geological Society Special Publication* No. 8.

Rickards, R. B., Rigby, S. and Harris, J. H., 1990. Graptolite biogeography: recent progress, future hopes. *Geological Society Special Publication.*

CHAPTER FIVE

Bulman, O. M. B., 1964. The Lower Palaeozoic Plankton. *Quarterly Journal of the Geological Society of London*, 120, 455–476.

Erdtmann, B. D., 1976. Ecostratigraphy of Ordovician graptoloids. *In* Bassett, M. G. (ed). *The Ordovician System: proceedings of a Palaeontological Association symposium.* 621–643. *University of Wales Press and National Museum of Wales, Cardiff.*

Finney, S. C., 1984. Biogeography of Ordovician graptolites in the Southern Appalachians. *In* Bruton, D. L. (ed.), Aspects of the Ordovician system. 167–176. *Palaeontological Contributions from the University of Oslo*, no. 295. *Universitetsforlaget.*

Fortey, R. A. and Bell, A. 1987. Branching geometry and function of multiramous graptoloids. *Palaeobiology*, 13, 1–19.

Fortey, R. A. and Cocks, L. R. M., 1988. Arenig to Llandovery faunal distributions in the Caledonides. *In* Harris, A. L., and Fettes, D. J. (eds), The Caledonian-Appalachian Orogen, 233–246. *Geological Society Special Publication*, No. 38.

Kirk, N., 1969. Some thought on the ecology, mode of life, and evolution of the Graptolithina. *Proceedings to the Geological Society of London*, 1659, 273–292.

Lapworth, C., 1897. Die Lebersweise der Graptolithen. *In* Walther, J. *Uber die Leberszweise fossiler meeresthiere. Zeitschrift deutsch geologisches Geselschaft*, 49, 238–258.

Mitchell, C. E., and Carle, K. J., 1986. The nematularium of *Pseudoclimacograptus scharenbergi* (Lapworth) and its secretion. *Palaeontology*, 29, 373–390.

Rickards, R. B., 1975. Palaeoecology of the Graptolithina, an extinct class of the Phylum Hemichordata. *Biological Reviews*, 50, 397–436.

Rickards, R. B. and Chapman, A. J., *in press*. The Bendigonian Graptolites (Hemichordata) of Victoria. *Memoirs of the National Museum, Victoria.*

Rickards, R. B., Rigby, S. and Harris, J. H., 1990. Graptolite biogeography: recent progress, future hopes. *Geological Society Special Publication.*

Skevington, D., 1973. Ordovician graptolites. *In* Hallam, A. (ed.), *Atlas of Palaeobiogeography*, 27–35. Elsevier, Amsterdam.

CHAPTER SIX

Berry, W. B. N., 1960. Graptolite faunas of the Marathon region, West Texas. *University of Texas Bureau Economic geology publication*, 6005, 1–179.

Cooper, R. A. and Fortey, R. A., 1982. The Ordovician graptolites of Spitsbergen. *Bulletin of the British Museum (Natural History) (Geology)*, 36, 157–302.

Erdtmann, B. D., 1976. Ecostratigraphy of Ordovician graptoloids. *In* Bassett, M. G. (ed.), *The Ordovician System: proceedings of a Palaeontological Association symposium*, 621–643. University of Wales Press and National Museum of Wales, Cardiff.

Finney, S. C., 1984. Biogeography of Ordovician graptolites in the southern Appalachians. *In* Bruton, D. L. (ed.), Aspects of the Ordovician System. 161–176. *Palaeontological Contributions from the University of Oslo, No. 295. Universitetsforlaget.*

Rickards, R. B., and Chapman, A. J., *in press*. The Bendigonian Graptolites (Hemichordata) of Victoria. *Memoirs of the National Museum, Victoria.*

Rickards, R. B., Rigby, S. and Harris, J. H., 1990. Graptolite biogeography: recent progress, future hopes. *Geological Society Special Publication.*

Skevington, D., 1974. Ordovician graptolites. *In* Hallam, A. (ed.), *Atlas of Palaeobiogeography*, 27–35. Elsevier, Amsterdam.

Skevington, D., 1974. Controls influencing the composition and distribution of Ordovician graptolite faunal provinces. *In* Rickards, R. B., Jackson, D. E. and Hughes, C. P. (eds), Graptolite studies in honour of O. M. B. Bulman. *Special Papers in Palaeontology*, 13, 59–74.

CHAPTER SEVEN

Palmer, D. C., 1986. The monotypic 'population' accompanying the lectotype of *Saetograptus varians* (Wood 1900). *Geological Society of London, Special Publication* no. 20, 249–260.

Rickards, R. B., 1979. Graptolithina. *In* Fairbridge, R. W. and Jablonski, D. (eds), *The Encyclopedia of Paleontology.* Dowden, Hutchinson & Ross, Stroudsberg, 351–359.

CHAPTER EIGHT

Bouček, B., 1957. The Dendroid Graptolites of the Silurian of Bohemia. *Nakladatelstui ceskoslovenske Akademie Ved*, 1–294.

Bulman, O. M. B., 1964. The Lower Palaeozoic Plankton. *Quarterly Journal of the Geological Society of London*, 120, 455–476.

Lapworth, C., 1878. The Moffat Series. *Quarterly Journal of the Geological Society of London*, 34, 240–346.

Watkins, R. and Berry, N. B. N., 1977. Ecology of a late Silurian fauna of graptolites and associated organisms. *Lethaia*, 10, 267–286.

CHAPTER NINE

Bulman, O. M. B. and Rickards, R. B., 1966. A Revision of Wiman's Dendroid and Tuboid Graptolites. *Bulletin of the Geological Institute of the University of Uppsala*, 43, 1–72.

Kozlowski, R., 1948. Les Graptolithes et quelques nouveaux groupes d'animaux du Tremadoc de la Pologne. *Palaeontologica Polonica*, 3, 1–235.

Skevington, D., 1967. Probable instance of genetic polymorphism in the graptolites. *Nature, London*, 213, 810–812.

Stebbing, A. R. D., 1970. The Status and Ecology of *'Rhabdopleura compacta'* (Hemichordata) from Plymouth. *Journal of the Marine Biological Association of the U.K.*, 50, 209–221.

Stebbing, A. R. D. and Dilly, P. N., 1972. Some observations on the living *'Rhabdopleura compacta'* (Hemichordata). *Journal of the Marine Biological Association of the U.K.*, 52, 443–448.

Urbanek, A. and Jaanusson, V., 1974. Genetic polymorphism as evidence of inbreeding in graptoloids. *In* Rickards, R. B., Jackson, D. E. and Hughes, C. P. (eds), Graptolite Studies in honour of O. M. B. Bulman. *Special Papers in Palaeontology*, 13, 15–18.

CHAPTER TEN

Bouček, B., 1957. The dendroid graptolites of the Silurian of Bohemia. *Rozpravy Ustredniko Ustavo Geologie*, 23, 1–294.

Bulman, O. M. B., 1950. Rejuvenation in a rhabdosome of *Dictyonema flabelliforme*. *Geological Magazine*, 87, 351–352.

Bulman, O. M. B., 1970. Graptolithina. *In* Teichert, C. (ed.), *Treatise on Invertebrate Palaeontology, Part V* (2nd edn.), Geological Society of America and University of Kansas Press, Boulder, Colorado and Lawrence, Kansas. i–xxxii 163pp.

Eichwald, E., 1842. Ueber das silurische Schichtensystem in Estland. *Zeitschrift fur Naturkinde und Heilk. der mediz. Akad. Zu st. Petersburg*, 1 and 2.

Goeppert, H. R., 1859. Die Fossile Flora der Silurischen, Devonischen, und unteren Kohlen-formation des sogenannten Uebergangsgebirges. *Nov. Acta Acad. Leopold. Carol.*, 27, Jena.

Hall, J., 1852. *Paleontology of New York*, 2, Albany.

Hall, J., 1865. Graptolites of the Quebec Group. *Geological Survey of Canada*, Canadian Organic Remains, decade 2, 1–151.

Hisinger, W., 1857. *Lethaea Suecica, seu Petrificata Sueciae*. Stockholm.

Holm, G., 1880. Gotlanders graptoliter. *Bihang till K. svenska Vet. Akademie*, 16, 1–32.

Kozlowski, R., 1948. Les graptolithes et quelques nouveaux groupes d'animaux du Tremadoc de la Pologne. *Palaeontologica Polonica*, 2, 1–235.

Nicholson, H. A., 1872. *Monograph of the British Graptolitidae*. Blackwood, Edinburgh and London. 133pp.

Ruedemann, R., 1947. Graptolites of North America. *Memoirs of the Geological of Society America*, 19, 652pp., 92 pls.

Salter, J. W., 1866. *In* Ramsay, The Geology of North Wales. *Memoirs of the Geological Survey of Great Britain.*

Von Brommell, 1727. Lithographiae Suecanae. *Acta literaria Sueciae Upsaliae* 1.

Wiman, C., 1895. Ueber die Graptolithen. *Geological Institute Uppsala*, 3, 239–316.

CHAPTER ELEVEN

Bulman, O. M. B., 1932–36. On the graptolites prepared by Holm. *Arkives für Zoologie*, 24a, 26a, 28a, 1–46, 1–29, 1–52, and 1–107.

Bulman, O. M. B., 1931. South American graptolites with special reference to the Nordenskiold Collection. *Arkives für Zoologie*, 22a, 1–111.

Bulman, O. M. B., 1944–1947. A monograph of the Caradoc (Balclatchie) graptolites from limestone in Laggan Burn, Ayrshire. *Palaeontographical Society (Monograph)*, 1–3; 1–42; 42–58; i–xi, 59–78.

Cox, I., 1933. On *Climacograptus inuiti* sp. nov. and its development *Geological Magazine*, 70, 3–19.

Holm, G., 1890. Gotlands Graptoliter. *Svenska Vetenskaps Akademie*, 16, 1–34.
Lapworth, C. *In* Elles, G. L. and Wood, E. M. R., 1901–18. Monograph of British Graptolites. *Palaeontographical Society (Monograph)*, 1–359.
Kraft, P., Ontogenetische Entwicklung und Biologie von *Diplograptus* und *Monograptus*. *Palaeontologische Zeitschrift*, 7, 207–249.
Kozlowski, R., 1948. Les Graptolithes et quelques nouveaux groupes d'animaux du Tremadoc de la Pologne. *Palaeontologia Polonica*, 3, 1–235.
Wiman, C., 1896. The structure of Graptolites. *Natural Science*, vol. 9, London.

CHAPTER TWELVE

Bulman, O. M. B., 1970. Graptolithina with sections on Enteropneusta and Pterobranchia *In* Teichert, C. (ed.), *Treatise on Invertebrate Paleontology*, pt. V (*2nd Edition*), *Geological Society of America and University of Kansas Press*, Boulder, Colorado and Lawrence, Kansas, l–xxxii, 1–163.
Fortey, R. A. and Cooper, R. A., 1986. A phylogenetic classification of the graptoloids. *Palaeontology*, 29, 631–654.
Hall, J., 1865. Graptolites of the Quebec Group. *Geological Survey of Canada, Canadian Organic Remains*, dec.2, 1–151.
Koren', T. N. and Rickards, R. B. 1979. Extinction of the graptolites. *In* Harris, A. L., Holland, C. H. and Leake, B. E. (eds), *The Caledonides of the British Isles – reviewed*. 457–466. *Geological Society Special Publication* No. 8.
Lapworth, C., 1873. On an improved classification of the Rhabdophorida. *Geological Magazine*, 10, 500–504, 555–560.
Mierzejewski, P., 1986. Ultrastructure, taxonomy and affinities of some Ordovician and Silurian organic microfossils. *Palaeontologica Polonica*, 47, 129–220.
Mu En-zhi, 1987. Graptolite Taxonomy and Classification. *Bulletin of the Geological Society of Denmark*, 35, 203–207.
Rickards, R. B., 1979. Graptolithina. *In* Fairbridge, R. W. and Jablonski, D. (eds), *The Encyclopedia of Paleontology*, Dowden, Hutchinson, and Ross Inc. Stroudsberg, 351–359.
Rickards, R. B., 1979. Early Evolution of Graptolites and related Groups. *Systematics Association Special Volume no.* 12, 435–442.
Rickards, R. B., Baillie, P. and Jago, G. *in press*. Upper Cambrian Dendroids from Smithton, Northern Tasmania. *Alcheringa*.
Rickards, R. B., Hutt, J. E. and Berry, W. B. N., 1976. Evolution of the Silurian and Devonian Graptoloids. *Bulletin of the British Museum (Natural History)* 28, 1–120.
Rickards, R. B., 1988. Anachronistic, heraldic and echoic evolution: new patterns revealed by extinct planktonic hemichordates. *Systematics Association Special Volume*, no. 34, 211–230.
Rigby, J., 1986. A critique of graptolite classification, and a revision of the suborders Diplograptina and Monograptina. *In* Hughes, C. P., Rickards, R. B. and Chapman, A. J., Palaeoecology and Biostratigraphy of Graptolites, 1–12. *Geological Society Special Publication*, no. 20.

CHAPTER THIRTEEN

Hall, J., 1865. Graptolites of the Quebec Group. *Geological Survey of Canada. Canadian Organic Remains* dec. 2, 1–151.
Lapworth, C., 1870. On the Lower Silurian Rocks of Galashiels. *Geological Magazine*, (1) 7, 204–209, 279–284.
Lapworth, C., 1873. On an improved classification of the Rhabdophorida. *Geological Magazine*, (1) 10, 500–504, 555–560.
Lapworth, C., 1878. The Moffat Series. *Quarterly Journal of the Geological Society of London*, 34, 240–346.
Lapworth, C., 1879. On the Tripartite Classification of the Lower Palaeozoic Rocks. *Geological Magazine*, (2) 6, 1–15.
Lapworth, C., 1879–80. On the Geological Distribution of the Rhabdophorida. *Annals and Magazines of Natural History*, 3, (1879), 245–257, 449–455; 4, (1879), 331–341, 423–431; 5, (1880), 45–62, 273–285, 358–369; 6, (1880), 16–29, 185–207.

CHAPTER FOURTEEN

Bates, D. E. B., 1986. The density of graptoloid skeletal tissue, and its implication for the volume and density of the soft tissue. *Lethaia*, 20, 149–156.
Bates, D. E. B. and Kirk, N. H., 1984. Autecology of Silurian Graptoloids. *Special Papers in Palaeontology*, 32, 121–139.
Bohlin, B., 1950. The Affinities of the Graptolites. *Bulletin of the Geological Institute of the University of Uppsala*, 34, 107–113.
Bulman, O. M. B., 1964. Lower Palaeozoic Plankton. *Quarterly Journal of the Geological Society of London*, 120, 455–76.
Garratt, M. J. and Rickards, R. B., 1983. Graptolite biostratigraphy of early land plants from Victoria, Australia. *Proceedings of the Yorkshire Geological Society*, 44, 377–384.
Garratt, M. J. and Rickards, R. B., 1987. Přídolí (Silurian) Graptolites in association with *Baragwanathia* (Lycophytina). *Bulletin of the Geological Society of Denmark*, 35, 135–179.
Kovacs-Endrödy, E., 1987. The earliest known vascular plant, or possible vascular plants in the flora of the Lower Silurian Cedarberg Formation, Table Mountain Graph, South Africa. *Annals of the Geological Survey of South Africa*, 20, 93–118.
Mitchell, C. E. and Carle, K. J., 1986. The Nematularium of *Pseudoclimacograptus scharenbergi* (Lapworth) and is secretion. *Palaeontology*, 29, 373–390.
Rickards, R. B., 1975. Palaeoecology of the Graptolithina, an extinct class of the phylum Hemichordata. *Biological Reviews*, 50, 297–436.
Rickards, R. B. and Garratt, M. J., 1990. Přídolí/graptolites from the Humervale Formation at Ghin Ghin and Cheviot, Victoria, Australia. *Proceedings of the Yorkshire Geological Society.*
Rigby, S. and Rickards, R. B., 1990. New evidence for the life habit of graptoloids from physical modelling. *Paleobiology.*
Theron, J. N., Rickards, R. B. and Aldridge, R. J., 1990. Bedding plane assemblages of *Promissum pulchrum*, a new giant Ashgill conodont from the Table Mountain Group, South Africa. *Palaeontology*, 33, 577–594.

APPENDIX 1

Bates, R. L. and Jackson, J.A. (eds), 1987. *Glossary of Geology*, (3rd edition). American Geological Institute, Alexandria, Virginia, 788pp.

APPENDIX 5

Black, R. M., 1988. *The Elements of Palaeontology*. Cambridge University Press, xii, 404pp.

Boardman, R. S., Cheetham, A. H. and Rowell, A. J., 1987. *Fossil Invertebrates*. Blackwell Scientific Publications, Palo Alto and Oxford, i–xi, 713pp.

British Palaeozoic Fossils 4th edn. 1975. British Museum (Natural History) i–vi, 203pp.

Bulman, O. M. B., 1927–1967. *British Dendroid Graptolites*. Palaeontographical Society Monograph, i–lxiv, 97pp.

Bulman, O. M. B., 1970. Graptolithina. *In* Teichert, C. (ed.), *Treatise on Invertebrate Palaeontology*, Part V (end edition, revised). The Geological Society of America and the University of Kansas, Boulder, Colorado and Lawrence, Kansas, i–xiii 163pp.

Clarkson, E. N. K., 1986. *Invertebrate Palaeontology and Evolution*. 2nd edition i–xiii Allen and Unwin, London 382pp.

Cleevely, R. J., 1983. *World Palaeontological Collections*. British Museum (Natural History) and Mansell Publishing Ltd. 365pp.

Elles, G. L. and Wood, E. M. R., 1901–1918. *Monograph of British Graptolites*. Parts I–XI, Palaeontographical Society Monograph, i–clxxi, 539pp.

Hancock, E. G. and Pettit, C. W. (eds.), 1981. *Register of Natural Sciences Collections in North West England*. Manchester Museum.

Murray, J. W. (ed.), 1985. *Atlas of Invertebrate Macrofossils*. Longman and Palaeontological Association i–xiii 241pp.

Nudds, J. and Palmer, D. 1990. Societies, Organizations, Journals and Collections *in* Briggs, D. E. G. and Crowther, P. C. (eds), *Palaeobiology: a synthesis*, Blackwell Scientific Publications, Oxford, pp.522–536.

Stace, H. E., Pettit, C. W. A. and Waterston, C. D., 1987. *Natural Science Collections in Scotland (Botany, Geology, Zoology)*. National Museums of Scotland i–xv 373pp.

Appendix 7

WHO CAN HELP ME?

The following is a list of BIG G members who have agreed to act as counsellors for any *bona fide* enquiries about graptolites that arise from reading this book or are not answered by the present text. The list is arranged geographically and in alphabetical order. So if you want some help with your graptolitic quest write to your nearest BIG G contact or to BIG G, Sedgwick Museum, Dept. Earth Sciences, Downing Street, Cambridge CB2 3EQ and we will either pass on your request to your local contact or endeavour to answer it ourselves.

Aberdeen
Dr Isles Strachan, c/o Department of Geology and Mineralogy, Marischal College, Aberdeen AB9 1AS.

Aberystwyth
Dr Dennis Bates, Dr David Loydell, Department of Geology, University of Wales, Llandinam Building, Aberystwyth, Dyfed SY27 3DB, Wales.
Dr Richard Cave, British Geological Survey, Regional Office for Wales, Bryn Eithyn Hall, Llanfarian, Aberystwyth, Dyfed SY 23 4BY, Wales.

Bristol
Dr Peter R. Crowther, City of Bristol Museum and Art Gallery, Queen's Road, Bristol BS8 1RL.

Cambridge
Dr Gwynn Morris, Dr Barrie Rickards, Dr Sue Rigby, Dr Margaret Sudbury, c/o Department of Earth Sciences, Downing Street, Cambridge CB2 3EQ.

Cardiff
Dr Douglas Palmer, Geology Department, National Museum of Wales, Cardiff CF1 3NP.

Essex
Dr Roger Hewitt (contact through BIG G Cambridge).

Glasgow
Dr David Skevington (contact through BIG G Cambridge).
Dr Elizabeth Bull (contact through BIG G Cambridge).

Keyworth, Nottinghamshire
Dr Adrian Rushton, Dr Steve Tunnicliff, Dr Dennis White, Dr Jan Zalasiewicz, British Geological Survey, Keyworth, Notts. NG12 5GG.

Leicestershire
Dr Mike Howe (contact through BIG G Cambridge).

London
Dr Richard Fortey, Department of Palaeontology, British Museum (Natural History), Cromwell Road, South Kensington, London SW7 5BD.

Newbury (Berkshire)
Dr A. Neil McLaurin, Rio Tinto Minerals Development Ltd., 4–8, The Broadway, Newbury, Berks. RS13 1BA.

Newcastle-upon-Tyne
Dr Richard Hughes, British Geological Survey, Windsor Court, Windsor Terrace, Newcastle-upon-Tyne, NE2 4HB.

Sheffield
Dr Martin White, Department of Geology, The University, Mappin Street, Sheffield, S1 3JD.

Wigan
Ms Amanda Chapman (contact through BIG G Cambridge).

Yorkshire
Dr Margaret Sudbury (contact through BIG G Cambridge).

Appendix 8

WHO WERE THE FAMOUS GRAPTOLITE WORKERS?

JOACHIM BARRANDE (1799–1883)

Barrande was a French engineer who made his home in Bohemia. In 1833, while supervising a road-cutting, some Cambrian fossils fired his interest and he began to amass a great collection of fossils of all sorts from the Palaeozoic rocks of the Prague district. He paid workmen to collect for him at promising localities and purchased choice material from amateur collectors. He resolved on a monumental undertaking – to write a description of all the Silurian fossils of Bohemia.

He began this endeavour with a magnificent work on trilobites, which appeared in 1852 and became a classic, but he encountered delays in the preparation of the lithographic figures. So in the meantime he wrote a comparatively modest work on the Graptolites of Bohemia (1850). In this book he not only described and most beautifully illustrated the graptolite species known from Bohemia, but, with his characteristic thoroughness, he reviewed all that was known about graptolites down to that time, summarising ideas on their structure, growth, distribution and mode of life. In consequence his work is the first text-book on the group and it became the prime reference and set the standard for subsequent workers. In one particular respect Barrande's ideas were badly mistaken: he held that the narrow end of the graptolite was the 'growing point' – a view overturned by the identification of the sicula as the first-formed zooid.

Barrande intended to return to the study of graptolites later in his life, but the enormous task of description he had set himself was held up when his pet theory of 'colonies' was attacked. Barrande devised this theory to account for occurrences of Silurian graptolitic shale in the midst of what we would call Ordovician shelly sandstone beds. His theory was essentially environmental; he thought that when conditions were suitable the graptolites came from elsewhere and colonised the Prague area; during the Ordovician these favourable periods were brief, but, he supposed, in the Silurian they were more extended. It also appeared that the graptolites of the Ordovician colonies were the same species as those of the Silurian. Other geologists, however, suggested that the strata containing the graptolite 'colonies' were introduced into the Ordovician by folds and faults. Barrande disliked being criticised and brought every argument he could muster to refute their views, writing four substantial pamphlets, 'Defense de colonies' I–IV; but the results of Lapworth's work in Scotland presented insuperable difficulties to the theory of colonies, though Barrande, immoveable, was preparing 'Defense de colonies V' at the time of his death.

Barrande left his great descriptive task unfinished, but his will provided for its continuation by others; in 1895 Jaroslav Perner prepared a monograph of the Bohemian graptolites uniform with Barrande's monographs of other fossil groups.

JAMES HALL (1811–1898)

In a review published in 1953, Prof. Bulman described Hall's *Graptolites of the Quebec Group* of 1865 as 'the first [graptolite] work of real understanding and insight' – this though Hall was no specialist in the group and his graptolitic work was (not least in bulk) overshadowed by his monumental contributions to the description of North American shelly faunas.

Hall was the most prolific, ambitious and successful palaeontologist in North America. His life's work for the New York State Geological Survey was conducted from his palaeontological power-house at Albany, N.Y., but, as he was constantly struggling for money to sustain his work and fund his publications, he also took contracts with other state surveys to give palaeontological advice and prepare descriptions and illustrations of their fossil faunas. Hall had described some graptolites from upstate New York, so when William Logan of the Canadian Geological Survey needed an assessment of the rich graptolite faunas from Levis, Quebec, that he had discovered in 1854, it was natural for him to turn to Hall. Hall started work on the fauna and the choice specimens were drawn with great skill by Hall's newly appointed assistant Robert Whitfield, but one of the many crises in Hall's turbulent career held up publication for ten years.

When the book appeared it was characteristic of its author's work – large in size, wide in scope, polished in presentation. Hall reviewed and illustrated general features of the graptolites and observed many points that had not previously been noticed, before embarking on a formal description of the Quebec species. These, being representative of an early Ordovician fauna, included several large, many-branched forms, and Hall surmised that the various species of *Monograptus* that Barrande had described from the Llandovery rocks of Bohemia were merely broken fragments of the sorts of many-branched species that he was describing from Quebec. Lapworth later proved this surmise incorrect when he demonstrated the fundamental importance of the sicula in his descrip-tion of Scottish Monograptids. Nevertheless, Hall's Graptolites of the Quebec Group, with its elegant illustrations, remains an important reference for the study of early Ordovician graptolites.

CHARLES LAPWORTH (1842–1920)

FOUNDER OF BRITISH GRAPTOLITE BIOSTRATIGRAPHY

Charles Lapworth was born on September 20th, 1842, at Faringdon in Berkshire. From an early age he was keen to follow a career in teaching, and in 1864 he graduated from the Training College at Culham, Oxford, with a first class certificate. Lapworth was fascinated with the literary works of Sir Walter Scott, and because of Scotts' association with the Scottish Borderlands, Lapworth took his first teaching post in Galashiels.

He had a history of poor health, and his doctor advised him to pursue interests in the open air. He became interested in geology, and began his long and distinguished association with the geology of the Southern Uplands by making a detailed study of the area around his home town of Galashiels, published in 1870.

It is clear that even at this early stage in his geological career, Lapworth recognized the possibility of using graptolites as stratigraphical tools. By 1872 Lapworth had moved the focus of his attention to the Moffat area, and the results of many years work here were published in 1878 in the classic paper entitled 'The Moffat Series'. The great achievements of this work were two-fold. By establishing for the first time the detailed usefulness of graptolites he was able to re-interpret the complex structure of the Southern Uplands.

The key to his success lay in his observations that the graptolites occur in a unique stratigraphical order. These observations led Lapworth to establish a sequence of graptolite zones which remain largely unchanged to the present day, and which continue to be used to solve stratigraphical problems in Ordovician and Silurian rocks throughtout the world. His work struck a resounding blow against Barrande's theory of colonies. Lapworth's 'typical section', at Dob's Linn east of Moffat, has recently been designated the international stratotype for the Ordovician-Silurian boundary. Shortly after 'The Moffat Series', Lapworth published the results of many years' research into the stratigraphical distribution of graptolites within and beyond the Southern Uplands.

Lapworth not only refined the stratigraphical use of graptolites, but also made many advances in understanding the palaeobiology of the group. Notable amongst these is his discovery (1873) that the sicula is the earliest growth stage from which the graptolite colony develops. The culmination of Lapworth's association with graptolite research came in his capacity as editor to 'A Monograph of British Graptolites' (Elles and Wood, 1901–1918). This work, with its range chart showing the stratigraphical distribution of the described species, remains a standard reference work for graptolite workers to the present day.

Lapworth's geological career was not limited to graptolite work. Most notably, he made a major contribution to the understanding of the geology of the North-west Highlands of Scotland, and ended the long and sometimes acrimonious dispute between Sedgwick and Murchison by introducing the Ordovician System (1879) to the Lower Palaeozoic. For this and other geological work Lapworth received numerous awards and university honours, and was elected a Fellow of the Royal Society in 1888. He retired in 1913 as Professor of Geology and Physiography at the University of Birmingham.

Charles Lapworth died on March 13th, 1920, in his seventy-eighth year.

GERHARD HOLM (1853–1926)

Early in his life Holm showed an interest in natural history, and one of his achievements in later life was to make a vital step towards the understanding of the natural history of graptolites by showing how their anatomy could be studied.

Holm was born and educated in Stockholm. After graduating in 1879 he moved to the University of Uppsala and in 1887 he moved to the Swedish Geological Survey. Throughout this period he made an intense study of various fossil groups and in the period up to 1901 produced over forty publications, many of them major works on a surprising range of topics. A feature of Holm's work was his skill in preparing fossils for illustration by extracting them from the rock. He pioneered the technique for extracting eurypterids (a kind of sea-scorpion) and graptolites, and published his sensational illustrations of isolated *Retiolites* in 1890. When in 1895 he published his next paper on isolated graptolites its importance was soon recognized, and two British students – Miss Elles and Miss Wood – immediately published an English translation of Holm's paper in order to make his results more widely available. Holm in the meantime extended his work on isolating graptolites to as wide a range of species as he could, and amassed a huge collection. But in 1901 Holm became the curator of the Palaeontological Department of the Swedish Natural History Museum and he never found time thereafter to work on the specimens that he had so carefully prepared. After his retirement in 1922 his health failed and he was unable to return to his former studies.

Holm's success depended on two factors. One was his skill in treating the rock and manipulating the fragile specimens so delicately that even the fine spines are intact; the other lay in his illustration, for which he had the help of the masterly illustrator B. Liljevall. He photographed the specimens and Liljevall then retouched the prints, producing figures of admirable fidelity and clarity. Over the years Holm and Liljevall produced many plates that were printed and ready for publication, but still awaited Holm's descriptive text, which unfortunately was never written. A few years after Holm's death, however, Oliver Bulman undertook the task of describing the material Holm had left, using the printed plates, and thereby he not only made it and Liljevall's figures known but greatly advanced knowledge of graptolite anatomy. Holm's career overlapped with that of a fellow Swede, Carl Wiman, a man perhaps not so widely known outside graptolite research circles, but he used similar techniques to Holm and also specialised in illustrations. One of his specialities was in careful serial sectioning of dendroid graptolites, thus interpreting internal structures with accuracy.

GERTRUDE LILIAN ELLES (1872–1960)

Gertrude Lilian Elles became a Cambridge Scholar and undergraduate in 1891 and read for the Natural Sciences Tripos. Both she and her contemporary, Ethel Wood, graduated with Firsts in Geology in 1895. After a year's research in Sweden with Holm, Wiman and Tornquist, Miss Elles returned to Cambridge and was to spend all the rest of her life there. Meanwhile Miss Wood had become assistant to Professor Lapworth in Birmingham and so began their collaboration of over 15 years which produced the Monograph of British Graptolites. Lapworth acted as editor, with Miss Wood preparing the illustrations and Miss Elles the text.

Miss Elles enjoyed field work and was a keen stratigrapher, producing some important papers on the geology of Wales and the Welsh borders at the same time as working on the Monograph. Later she wrote a paper on the graptolite faunas of the British Isles which gave a clear summary of the value of graptolites to the Palaeozoic stratigrapher.

Apart from her academic work she enjoyed music and was an active sportswoman, coaching teams for her college, Newnham, where she later became a Fellow. Her interest in the College and its welfare was constant. She also worked hard for the Red Cross, commanding a hospital in Cambridge during the First World War, for which she received the MBE.

In the academic field she was one of the first woman admitted to the Geological Society and later became the first woman Reader of the University. She was a D.Sc. of Trinity College Dublin and an Sc.D. of Cambridge.

In her late seventies she had become very deaf and her short stocky figure, clad always in a thick tweed suit whatever the weather, caused some amusement as she progressed slowly, puffing slightly and aided by a stick, along the corridors of the Sedgwick Museum. She would go up in the lift, accompanied by Henry Brand, a Sedgwick Museum lab technician to work it for her, to her large attic room. There books, papers and specimens lay everywhere – 'stratified just like a sedimentary rock' – as A. G. Brighton, the Museum curator described when he later sorted its contents. She still supervised first-year women undergraduates such as myself, who hardly then knew what graptolites were, but she would sometimes show us specimens and her microscope was always kept on the bench under the window as she insisted that natural light was so much better than a lamp.

Although by this time 'G.L.E.' was very much a Cambridge 'character' and very isolated by her deafness, her clarity of mind and kindly nature remained and she was regarded with much affection both by her friends at Newnham and her colleagues in the geological world. She died in 1960 at the age of 88.

ROMAN KOZLOWSKI (1889–1977)

Roman Kozlowski understood that scientific knowledge can be a fragile thing. As the second world war erupted around Poland, he had hidden a manuscript, the result of years of labour, in a basement of the Seismological Institute in Warsaw. The Institute was bombed, then looted. Miraculously, he found part of the manuscript in the rubble afterwards, the remainder being found later, buried under snow. Nothing survived of the original specimens or photographs – except that copies of the negatives had been sent to Paris just before the war broke out. During the Warsaw uprising, the manuscript was hidden again, in the central heating system of a house. The house was destroyed; but the manuscript somehow survived inside its improvised shelter, and was again rescued. When published after the war, with the help of the Parisian negatives, a major palaeontological riddle was solved, and a great scientific reputation established.

For Kozlowski, this was perhaps the most satisfying moment in a lifetime of scientific work. Born in Wloclawek, north-west of Warsaw, in 1889, he was educated at Frieburg and the Sorbonne. In his twenties and early thirties, he worked in Bolivia, as Professor of Geology and Mineralogy, and later as Director, of the School of Mines. His ingenuity in examining and describing fossil material was already in evidence. His early work was on brachiopods, and the detail in which he monographed the complex internal skeletons of the specimens he collected in Bolivia influenced much subsequent research into this fossil group.

Back in Warsaw, a Professor of a steadily expanding Department of Geology, he made a discovery that changed the course of his career. He noticed that certain fine-grained siliceous rocks (cherts) of Upper Cambrian age in the Holy Cross mountains had the remains of early many-branched graptolites entombed within them, perfectly preserved, like flies in amber (see p. 48). He was able to extract these fossils by dissolving the chert in acid, and study them just as one would study modern biological specimens. His precise, painstaking observations of these early graptolites, that barely survived the second world war, solved a question that palaeontologists had been asking for more than a century – just what *were* graptolites?

Previously, it had been thought that they were related to corals, or bryozoans. Kozlowski showed that they were neither. Rather, they were related to the pterobranchs, within the phylum Hemichordata (see Chapters 1 and 12), and so, among other things, were much more closely related to *us*. This conclusion has since been elaborated, but rarely seriously disputed.

In later years, despite fame and international honours, he simply carried on working as assiduously as before. He established a school of palaeontology in Poland (still thriving), established and edited palaeontological journals, and conducted further detailed studies into the remains of graptolites (and other animals), acid-etched out of many tons of rock. He died in 1977, at the age of 88, having worked almost to the end.

OLIVER MEREDITH BOONE BULMAN (1902–1974)

Oliver Bulman was one of those whose childhood hobby of collecting minerals and fossils developed into an interest in geology while still at school. He went on to study geology at Imperial College, London, and after graduating with first class honours worked on a joint Ph.D. thesis with his life-long friend C. J. Stubblefield. The subject was a study of the Shineton Shales of the Wrekin area, and as well as sharing in the mapping and structural work, Bulman took responsibility for the study of the *Dictyonema* species, among other groups. Thus his interest in graptolites began with the dendroids. Later he spent two years in Cambridge working under the supervision of Miss G. L. Elles and this research was published as the first two parts of a monograph on British Dendroid Graptolites, a work which was only completed in 1967 after he had resigned his professorship a year before retirement age so as to devote more time to research.

After several years teaching zoology and geology at Imperial College, he returned to Cambridge in 1931 as a Demonstrator in Geology, successively becoming Lecturer, Reader in Palaeozoology, and Woodwardian Professor of Geology (1955–66). His interests extended to all graptolites and he wrote a large number of papers during the following years. The work on which these were based showed his mastery of techniques – delicate work with a needle to remove flakes of covering rock; isolation with acid and mounting in Canada Balsam: serial sectioning to produce splendid enlarged models. The results were communicated in a clear style and with effective and elegant illustrations which testify to his remarkable artistic ability. Later works surveyed the group as a whole from his wide knowledge of it and particularly valuable in this respect are the two editions (1955 and 1970) of the Graptolithina section of the Treatise of Invertebrate Paleontology.

Tall and upright, slim and despectacled, he appeared a somewhat austere figure to the undergraduates to whom he lectured on palaeontology, but all admired his clarity of expression and especially the stylish drawings and diagrams which he executed with such ease on the blackboard. When one knew him better he seemed less remote and one could appreciate his dry wit in the coffee room and his friendly but unobtrusive guidance and support for one's research efforts. His large, rather dark room in the Sedgwick Museum was dominated by a long table in the exact centre, covered with orderly trays of specimens and neat piles of books and papers. He presided at the far end of this, often with a cigarette, while he delicately manipulated a specimen or made a drawing from the binocular microscope with each objective and eyepiece engraved with the familiar initials 'O.M.B.B.' He was left-handed; his writing was extremely neat and regular.

His high standard of scholarship led many students to Cambridge to work under him, and he travelled widely to countries such as Sweden and Poland where there were classic graptolite localities, to collect, study and share his expertise. He edited the Geological Magazine for 38 years and held various offices including presidencies in the Geological Society and the Palaeontological Association. His many honours included election to the Royal Society at the age of 37. A collection of papers on graptolites by colleagues and former pupils was designed to form a Festschrift for him, but became instead a memorial volume published in 1974.

MU EN-ZHI (A.T.MU) (1917–1987)

Prof. Mu En-zhi (A.T. Mu) from Fengxian, Jiangsu Province, China, graduated from the Department of Geology, Geography and Meteorology of Southwest University, China in 1943. He was a member of the Department of Earth Sciences, Academia Sinica, the Deputy Director of the Nanjing Institute of Geology and Palaeontology, Academia Sinica, and the Palaeontological Society of China. He was also the titular member of the Silurian Subcommission and the representative for China in the Scientific Committee of IGCP.

Prof. Mu published 135 papers. In the late forties, he recognized the age of the Wufeng graptolite shale as Ashgill. In 1950, the same year Bulman's Treatise (First edition) was published, he proposed a new classification of the graptolites; and many of his taxa, such as Sinograptidae, Abrograptidae etc, have been recognized and used internationally. In this last 30 years, he and his students defined Chinese graptolite zones from the late Cambrian to early Devonian and described the main graptolite faunas in China. In the sixties, he defined 7 graptolite sequences through the Ordovician and Silurian. In the seventies he emphasized that ecological differentiation was the main cause of the graptolite faunal distributions. Prof. Mu also worked on echinoids and nautiloids, at least 14 papers being published.

As a stratigrapher, he reviewed the Silurian of China in 1962 and again in 1986. He joined and organized a series of stratigraphical working teams to investigate the early Palaeozoic rocks in a widespread area in China, including Northeast China (Taizihe area), the Qilian Mountains, the Yangtze area, and South China including Hainan Island. This work formed not only the foundation of early Palaeozoic stratigraphy in China, but also contributions to Chinese Petroleum exploration.

He contributed to Chinese geological education. From the fifties he was part-time Professor in the Department of Geology both in Peking and in Nanjing universities. His students now work in the Institutions, Universities, Museums and Geological Surveys in different parts of China.

Prof. Mu travelled widely outside China and was very active in the Silurian Subcommission and Ordovician Silurian Boundary Working Group.

Prof. Mu left a geologist family. His eldest son, Mu Xi-nan is an Associate Research Professor in the Nanjing Institute of Geology and Palaeontology; his son-in-law and daughter-in-law are both geologists. His wife, Huang Ti-fen, was a school teacher. She is fortunate to have managed the family in such a way that allowed her husband to concentrate on research.

FIGURE EXPLANATIONS

Note that the number/s in brackets following the fossil name refer to the classification and evolution text figures of appendices 3 and 4.

1. Leaf shaped in both name and appearance but actually a 475 million year old graptolite from Australia magnified ten times. *Phyllograptus nobilis* (Phyllograptidae, appendices 3 and 4). Darriwilian, (Llanvirn) Ordovician; from Gibbo, Victoria, Australia; thecal tubes well seen to left and right, either side of an obscure axial region, which represents the 'head on' apertural view of two similar rows of thecae usually only seen in three dimensional material. This is a typical flattened specimen, white chloritic material on a bedding plane of black (graptolitic) shale. The relatively pointed proximal end (bottom of the photo) of the leaf-shaped colony, helps distinguish *Phyllograptus* from *Pseudophyllograptus* which is relatively broader at that point and has thecal tubes which grow horizontally rather than being curved downwards. Scale bar is 1mm. Photo Bill Berry, USA.

2. Many graptolites were adept geometricians, witness the outstretched branches of this large *Goniograptus* (7). Bendigonian (Arenig) Ordovician; from Campbelltown, Victoria, Australia; this large spectacular colony (rhabdosome) grows from the centre (proximal region), and the geometrical branching pattern seen is constant for each species. In horizontally disposed forms such as this, the thecae are not usually visible (because they face downwards or upwards into the sediment) but in this specimen the branches (stipes) have twisted sideways revealing the fret-saw edge typical of most graptolite stipes. These multibranched rhabdosomes are amongst the earliest of the graptolitic plankton. Scale bar is 5 mm.

3. Webbed arrangements of early branching in *Loganograptus rectus* (5). Bendigonian (Arenig) Ordovician; Sandon, Victoria, Australia; proximal region of a horizontally disposed multibranched colony; thecae not seen (facing down, or up, into the sediment) but the size and geometrical arrangement of stipes makes for species identification. Scale bar is 2 mm.

4. Many branches linked together form the inverted conical net of *Rhabdinopora parabola* (until recently termed *Dictyonema*, a name now restricted to those graptolites which lived attached to the substrate) (Anisograptidae in Text figure 1 and appendices 3 & 4). Tremadoc, early Ordovician; Newfoundland, Canada; the colony has a triangular outline which represents a flattened (narrower) cone which widened upon flattening to the bedding plane. The stipes are connected by thin rods called dissepiments. Thecae can only be seen faintly towards the edges of the colony, but the sicula, from which the whole colony grew, is just visible at the apex of the cone. This large colony is seen in profile, but the star-shaped colony just below it is the same species preserved in plan view: only early growth stages like this are likely to be preserved in plan, because as the cone grows longer

it would have been more usual for it to lie on its side on the sea bed, after death. Other specimens can be seen on different bedding planes, which is typical of black graptolitic shales. These are the earliest known planktonic graptolites. Scale bar is 2 mm. Photo B-D. Erdtmann.

5. Writing on the rocks! Hieroglyphics produced by abundant dichograptids (pendent three-stiped *Tetragraptus*, horizontal *Expansograptus* and *Sigmagraptus*). Here the original black shale has been completely bleached (see also Figs 112 and 116, for partial stages in this process) effecting a strong contrast between graptolites and matrix. Further bleaching, caused by weathering, would also eliminate the graptolites. Thecae are seen as fret-saw blade outlines except, top centre, where the stipes are viewed from the back or dorsal side, where no thecae are visible on the stipe. Scale bar is 1 mm.

6. The helical growth spiral of *Monograptus proteus* (4). Llandovery, Silurian, Germany; spirally coiled colony now flattened upon the bedding plane as a silvery chlorite film on black shale. The colony grew from the slim position near the bottom of the page, and it should be noted that in this case the thecae, whilst having the fret-saw outline referred to in Fig. 2, also have a hooked appearance. Further, the specimen is not only flattened, but *stretched*, in a direction from top right to bottom left as a result of the rock being compressed during folding by forces operating normal to that direction. Scale bar is 2 mm. Photo Manfred Schauer.

7. Not a graptolite but a hydroid polyp of *Antennularia*. Recent, prepared by Professor Oliver Bulman for direct comparison with the living *Rhabdopleura* zooids (Figs 8 and 9). (Graptolites and rhabdopleurans were probably closely related; see Ch. 1). Note the living tissue connection (stippled appearance) of the zooids, within the peridermal sheath, and the buds of zooids-to-be. Scale bar is 100 μm.

8, 9. Probably the nearest living relative of the graptolites, *Rhabdopleura compacta* (Pterobranchia of Appendix 3). Recent; respectively, a retracted zooid with its tube (coenecial tube, showing zig-zag growth increments) and a partially extended zooid showing lophophore, tentacles, and food grooves along the tentacles. Note similarity of Fig. 8 with Figs 33 (of *Rhabdopleura* zooids) and 34 (of graptolite zooids). Scale bar is 50 μm.

10, 11. *Sinograptus typicalis* (6), Llanvirn, Ordovician; Zhejing Province, China; two views of the same specimen illustrating the increased contrast with the matrix, achieved by wetting the rock. In the field, water can be used to improve visibility, but in the laboratory absolute alcohol is better. In this case glycerine was used, but it is difficult to remove air bubbles from such a wetting agent. The thecal tubes in this genus are multisigmoidal and their apertures obscure. The sicula, at the junction of the two stipes, is just visible. Scale bar is 0.5 mm. Photo Chen Xu, China.

12. Classic winged form with horizontally extended branches pivoted about a prominent sicula. *Expansograptus similis* (5). Arenig, Ordovician; Newfoundland, Canada; simple thecal tubes, arranged in two stipes which diverge from the original conical individual of the colony, the sicula. The sicula is the conical body sticking up on the dorsal (or anti-apertural) side of the stipes: its aperture is at the opposite end to the pointed apex and is turned slightly towards the left-hand series of thecae. There is no virgella ('sicular spine') but a broad process on the right of the aperture. The first *theca* of the colony originates

as a narrow tube, just visible, almost at the apex of the sicula. Note the very fine grain of the sediment in which the graptolite is entombed. Scale bar is 1 mm. Photo Henry Williams.

13. The starting point of all graptolite colonies; stage 1 – the prosicula, seen here in *Climacograptus typicalis* (2, 3). Caradoc, Ordovician; Oklahoma, USA; high magnifications of the original, tiny conical structure (sicula) common to all graptolites and usually of this shape, look like the nose cone of a rocket. It is in two basic parts: the upper, pointed, part with at least 10 (strengthening) threads along its length; and a lower part with closely spaced, regular, curved, growth lines which indicate incremental growth. The upper part is the earlier-formed, and is termed the *prosicula*, and it is characterised by longitudinal threads which converge at its apex, and a fainter spiral thread. The prosicula is secreted as a complete unit, only adding a few longitudinal threads. The lower part is the *metasicula*, better seen in Figs 14 and 15. This specimen has been chemically isolated from the matrix (details Ch. 11) but the features illustrated can also be seen on well preserved material in the rock. Scale bar is 50 μm.

14. The starting point of all graptolite colonies; stage 2 – the metasicula, seen here in *Climacograptus typicalis* (2, 3). Caradoc, Ordovician; Oklahoma, USA; metasicula as in Fig. 13, but showing how growth increments, called *fuselli* build and shape the whole sicular cone, including the virgellar spine (bottom left) and the two anti-virgellar spines (bottom right), typical of this species. The cowl to the left of the metasicula is the down-growing part of the first theca of the colony, also built of the same fusellar increments (as is the whole graptolite colony). This individual, therefore, died at a very immature stage of growth with only two zooids in the colony, and one with its house half built! Note also the fainter bandage-like markings along the length of the metasicula. These are cortical bandages, usually added to the outside of the colony for strengthening and shaping and smoothing purposes. They are biochemically the same as the fuselli (the protein collagen) and were probably secreted by the same zooidal organ from secretary cells on its surface. (Published with permission of the Palaeontological Association). Scale bar is 50 μm.

15. Budding begins; development of the early thecae from the sicula. This is the same species as in Fig. 14 but it shows the cortical bandages more clearly, the zig-zag junction of the fusellar growth increments (especially clear at the bottom of the page) and the origin of the first theca more clearly: the last grows from a pore resorbed in the metasicular wall (the pore can be seen through the transparent metasicula at its bottom left extremity on the photo). The origin of the virgella (see also Fig. 14) is also well shown all the way down the left of the metasicula where the growth increments dip downwards towards a dark band. Scale bar is 50 μm.

16. The long and the short; thecal dimorphism in a dendroid, *Dictyonema* sp. (Dendroidea in appendices 2 and 3). Ordovician; Sweden; a fragment of stipe from a bottom-living (benthonic) graptolite, exhibiting the universal fusellar growth increments. In this order there are two types of thecae, the conspicuous *autothecae* pointing apertural denticles to the left in this photograph; and the diminutive *bithecae*, the apertures of which are seen just below the autothecal apertures and slightly to the right. These may have housed, respectively, female and male zooids (see chs 1 and 9). The upper of the two bithecae

is on the side nearest the camera; the other is seen more as a dark shadow through the transparent autotheca. Scale bar is 100 μm.

17. A bizarre *Adelograptus* sp. (Anisograptidae of appendices 3 and 4). Ordovician; Newfoundland, Canada; with a somewhat unusual wide sicula, yet showing all the basic sicular characters, namely prosicula with strengthening rods and metasicula with fuselli. The first theca of the colony clearly seen originating on the prosicula (typically of early Ordovician graptolites). The theca themselves are highly unusual, being very long, thin tubes, though this feature is seen at intervals through the record (e.g. Fig. 55). Scale bar is 50 μm. Photo Henry Williams.

18. The upraised arms of *Isograptus subtilis* (9). Arenig, Ordovician; Newfoundland, Canada; showing much greater development of the nemal rod than on earlier photographs. Note that the stipes are almost back-to-back, but not quite; and that the early thecae grow strongly downwards, pendent, from an origin high on the sicula. It is exceedingly difficult, in forms still in the rock, to distinguish the sicula from early thecae. Note the sedimentary plugs in the apertures. Scale bar is 0.5 mm. Photo Henry Williams.

19. *Amplexograptus maxwelli* (2, 3). Middle Ordovician; Wisconsin, USA; early growth stage cleared with Schultz's Solution (Ch. 11); fusellar building blocks well seen, as are the cortical bandages. Cortical bandages often radiate, as here, from particular thecae, suggesting that a particular zooid was responsible for their secretion. The tangle of tubes at the bottom right is a series of proximal buds as the two thecal series of a scandent biserial become established (Fig. 22). Note the very oblique zig-zag sutures of the fuselli, at the extreme left of the uppermost theca. Scale bar is 50 μm.

20. *Paraclimacograptus innotatus* (2, 3). Llandovery, Silurian; Ural Mountains, USSR; fragment of periderm seen under the SEM showing individual bandages criss-crossing in the cortical layer, forming an external plastering on the colony walls. Each bandage shows fibres (called fibrils) parallel to their length. The fibrils are composed of the protein collagen. Very thin sheets often envelop individual bandages (top left, bottom right). Scale bar is 5 μm.

21. Wrapped up in bandages; *Climacograptus typicalis* (2, 3). Caradoc, Ordovician; USA; distal fragment of a colony with thick cortical layer, mosaic of SEM pictures; numerous cortical bandages visible, completely obscuring the underlying fusellar incremental layer; thecal tubes have a sharp knee-bend in them, called a geniculum, and this was grown out above the previous thecal aperture providing a roof above it; the result is a series of deep excavations along each side of the biserial colony; and from the upper part of these excavations – the geniculum – it is common to see a sharp flange developed, as seen here. Scale bar is 200 μm. (Published with permission of the Palaeontological Association).

22. Beneath the bandages; *Amplexograptus maxwelli* (2, 3), details as Fig. 19, the nema can be seen through the transparent lateral rhabdosomal wall. Each theca buds alternately across the colony, i.e. there is no median septum separating the left from the right series (it is said to be aseptate). However, there is a conspicuous septum between individual thecae, called the *interthecal* septum. Note how the fusellar growth rings shape the thecal apertures. Succeeding geniculate thecae (knee-shaped, Fig. 21) overgrow early thecae,

overhanging their apertures, providing them with a roof! Scale bar is 100 μm. (Published with permission of the Palaeontological Association).

23. Perfectly pyritised; *Petalograptus ovatoelongatus* (2, 3). Llandovery, Silurian; central Wales; three dimensional pyritised specimen preserved as an internal mould. Thecae are long, overlapping, simple tubes, slightly curved, with apertures facing away from the axis of the colony. Axis is marked by a median septum separating the two thecal series. The cross section of such a colony is rather tabular. Note that the *interthecal* septa do not reach the median septum: the small gap connecting each thecal tube at its base is the *common canal*. The sicula is visible at the proximal end, is not covered by growth of the early thecae, and this is, therefore, an *obverse* view. Scale bar is 0.5 mm.

24. *Cystograptus vesiculosus* (2, 3). Low Llandovery, Silurian; Germany; slab of black graptolitic shale with flattened specimens including all *growth stages* from sicula to fully mature adults. Early growth stages are the small, spiky parts (a very long sicula). Although the thecae look like simple tubes this is an artifact caused by severe flattening. The nema in these forms is greatly expanded to a three-vaned, dart-like structure, well seen on the mature specimen at the bottom left. It is likely that such vanes are to aid feeding rotation of the colony. When all growth stages are present, and the assemblage dominated by a single species, it is possibly the result of a single plankton kill, followed by gentle sinking to burial in the mud. Scale bar is 2.5 mm. Photo Manfred Schauer.

25. A contrast of preservations. *Climacograptus angulatus magnus* (2, 3). Llanvirn, Ordovician; Oslo, Norway; pyritised biserial scandent form, periderm having flaked away showing Fools' Gold reflecting beneath; thecal tubes are sigmoidal and the apertures facing slightly inwards rather than away from the colony; the two thecal series are separated by a median septum, a dark line down the centre of the colony. Note the knife marks in the black shale where obscuring matrix has been removed. The specimen to the left is *the same species*, but quite flattened on the bedding plane and presents a rather different appearance. The spike projecting 'downwards' from it is the nema: the specimen lies in the opposite direction to the pyritised specimen and its distal end is pointing down the page. Scale bar is 1 mm. Photo Bill Berry, USA.

26. A complete dendroid, *Dictyonema* (Dendroidea of appendices 3 and 4). Ludlow, Silurian; Orange, NSW, Australia; the conical rhabdosome is preserved with the open end of the cone apparently face down on the bedding plane but the way up of the slab is not known. The apex of the cone terminates in a basal holdfast, the dark patch where the stipes come together. This specimen occurs in limey sandstone and the holdfast may have been attached to a small pebble or shell fragment. Scale bar is 2.5 mm. Specimen, Tony Wright, Australia.

27. The knotted stipes and introverted thecae of a Chinese *Dicellograptus szechuanensis* (1). Ashgill, Ordovician; Guizhou Province, China; this species has two spirally intertwined stipes; a badly preserved proximal region with basal spines is visible bottom left; the thecae are strongly turned inwards (introverted) at their apertures; the black colour is because the original periderm is still present, the three dimensional stipe probably being infilled with pyrites. Scale bar is 1 mm. Photo Chen Xu.

28. Remarkable detail preserved despite the complete flattening and replacement by

chlorite of *Orthograptus calcaratus* (2, 3). Caradoc, Ordovician; Scotland; note the semicircular thickenings around the thecal apertures and the nemal rod running the length of the colonies. The preservation is classic silvery chloritic material on a black shale background. Scale bar is 2 mm.

29. A theca full of pyrite. *Monoclimacis* sp. (4). Wenlock, Silurian; Cornwallis Island, Canadian Arctic; view of a single geniculate thecal aperture (see Fig. 21 also) with infilling of pyrite crystals; pyrite tends to form internal moulds of the periderm, and occasionally external moulds, but it very rarely *replaces* the periderm and only on infrequent occasions does it disrupt the periderm. Scale bar is 25 μm.

30. Entombed in chlorite. *Climacograptus normalis* (2, 3). Llandovery, Silurian; Wales; specimen is completely pyritised, in three dimensions, and is encased in fibrous chlorite, the fibres radiating from the specimen; the fact that the graptolite is internally filled with pyrite before compaction of the sediment, and before tectonic deformation, allows the chlorite to grow in 'strain shadows'. Note the medium septum separating the two thecal series for almost their whole length; and the prominent, though short, virgellar spine at the proximal end. The sicula is partly visible, just above the virgella, but the first two thecae of the colony overgrow it. This is, therefore, the *reverse* view of the colony. Scale bar is 1 mm.

31. A look at the apertures. *Pseudoglyptograptus* sp. (2, 3). Llandovery, Silurian; Wales; pyritised specimen in subapertural view. A median septum is just visible to the left of the specimen. The thecae seem slightly sigmoidal and constricted, typical features of the genus, but in specimens flattened in this orientation all thecal details are totally lost. Scale bar is 1 mm.

32. *Pseudoplegmatograptus obesus* (4). Llandovery, Silurian; Canadian Arctic; proximal view of rhabdosome (see also Fig. 65) showing lists (thickened bars or rods) supporting an attenuated periderm (see Ch. 3 for details). Scale bar is 0.25 mm. Photo Michael Melchin.

33. Shy, retiring zooids of *Rhabdopleura compacta* (Pterobranchia of appendix 3). Recent. Coenecium (mass of creeping tubes) of colony showing retracted zooids (dark) and connecting black stolons, well seen in the lower part of the picture. Growth rings of the creeping tubes visible in places. Some immature zooids, in the early budding stage, may also be present. Note the great similarity to the graptolite zooids and stolons of Fig. 34. Scale bar is 250 μm.

34. Equally shy and retiring fossil zooids: the only known preserved graptolite zooids, of *Psigraptus jacksoni* (Anisograptidae of appendices 3 and 4). Tremadoc, Ordovician; Tasmania, Australia; these are the only graptolite zooids known – retracted zooids, dark coloured, attached to stolons; the murkier areas alongside them are the remains of the thecal tubes (in a different light angle they show fuselli). See *Psigraptus* also in Fig. 121. Scale bar is 500 μm.

35. 3D spiral revealed by *Monograptus turriculatus* (4). Llandovery, Silurian; Sweden; 3D SEM shows original helical spiral form of rhabdosome and the hooked thecae, although the thecal spines are mostly broken off (see Fig. 37). Scale bar is 500 μm.

36. Flattened spiral of *Monograptus proteus* (4). Llandovery, Silurian; central Wales; showing the effect of flattening upon a helically coiled spiral rhabdosome. Hook-shaped proximal thecae visible. The dorsal side of the stipe is marked by a black, nemal rod, and this clearly depicts the manner of the twisting. Scale bar is 1 mm.

37. Flattened and deformed spiral, but still with exquisite detail. *Monograptus turriculatus* (4). Llandovery, Silurian; Germany; whilst thecal details are obscure the abundant thecal spines are well preserved in this highly flattened specimen. Some minute hook-shaped rhabdosomes also occur on the same bedding plane. Scale bar is 2 mm. Photo Manfred Schauer.

38. Spiral vane for spiral motion through the water. *Petalograptus* sp. (2, 3). Llandovery, Silurian; Germany; the nema expands distally to become a twisted ribbon with thickened edges, almost certainly an expression of original spiral coiling. Contrast the three-vaned nemal structures of Fig. 24. Scale bar is 2 mm. Photo Manfred Schauer.

39. *Monograptus turriculatus* and *Monograptus marri* (4) details as Fig. 37; specimens are not only flattened but tectonically deformed, the compression being effectively from the sides of the page, producing *narrow* cones at right angles to this compression direction (top of page) and *wide* cones parallel to it (bottom of page). Scale bar is 4 mm. Photo Manfred Schauer.

40. Buried in sandstone. Sandstone/siltstone from Silurian of Morocco, with shell debris in the upper layers, and with many small rounded 'grains' surrounding them: the 'grains' are actually oriented graptolite rhabdosomes pointing into/out of the page. Right at the bottom of the sample is a specimen at right angles to that direction so that its length is visible. Scale bar is 4 mm.

41. 'Silver filigree' of *Rhabdinopora parabola* sp. (Anisograptidae of Text Fig. 1, and appendix 4); Tremadoc, Ordovician; Newfoundland, Canada; complete but flattened conical colony dependent from a sicula and with stipes united by frequent dissepiments; numerous fragments of the same species on the same and other bedding planes. One of the first planktonic graptoloids. Scale bar is 1 mm. Photo B-D. Erdtmann.

42, 43. The part and counterpart of *Aspidograptus implicatus* (Dendroidea of appendices 3 and 4). Arenig, Ordovician; Shropshire, England; notice regular dichotomies, parallel arrangement of stipes, but the total lack of dissepiments. The colony was probably fan shaped, growing from a holdfast where the stipes converge. The two figures are part and counterpart and exemplify how important it is to keep both: in this case almost all the periderm has gone with one part. Scale bar is 2 mm.

44. *Staurograptus dichotomus* (Anisograptidae of Text figure 1 and appendix 4). Tremadoc, Ordovician; Newfoundland, Canada; showing four stipes growing from origin followed by four equally spaced and rapid dichotomies; following the next dichotomy the stipes become supported by thin, black dissepimental rods; thecal profiles are visible in places. Unaltered black periderm of this type is often very fragile and may need strengthening by soaking in thin glue. Scale bar is 2 mm. Photo B-D. Erdtmann.

45, 46. Like forest stumps, the tiny living tubes of a 450 million year old colony of encrusting tuboids. *Discograptus schmidti* (Tuboidea of text figure 1, and appendix 3).

Middle Ordovician; Sweden; two autothecae showing ventral and dorsal apertural processes; and the basal disc with groups of autothecae rising from it; bithecae are tiny tubes, barely visible, and confined to the basal disc area of the specimen in Fig. 46. Scale bars are respectively 75 μm and 100 μm.

47, 48. 500 million years old; the beginning of the earliest graptolite plankton. Anisograptidae (indeterminate), Tremadoc, Ordovician; Shropshire, England; remarkably preserved specimens showing the prosicular origin of the first theca, the narrowing of the tip of the sicula to a *cauda* (central in Fig. 48) and the further narrowing of that to become the nemal rod or tube. Scale bars are respectively 50 μm and 125 μm.

49. Perhaps an attempt to lose weight! *Plectograptus* (*Sokolovograptus*) sp. (4). Wenlock, Silurian; Canadian Arctic; note that the network of lists (clathrial network) more or less indicates the positions of thecae and thecal apertures. This specimen is now preserved in a resin block, rather than in the usual medium which is glycerine. Scale bar is 500 μm.

50. *Amplexograptus maxwelli* (2, 3), same specimen as in Fig. 19; shows the early growth of thecal tubes and thecal budding, the virgella, pair of anti-virgellar spines, and a sub-apertural spine on th1. This whole proximal arrangement of spines is an Ordovician feature. Scale bar is 100 μm. (Published with permission of the Palaeontological Association).

51. The Pagoda style was invented by Chinese graptolites. *Pseudisograptus manubriatus janus* (8). Ordovician; China; three dimensional pyritised specimen showing complex arrangement of downward growing, rapidly budding early thecae. Note also that the early parts of all the thecae have a distinct fold (which would *not* be seen on flattened specimens – e.g. Fig. 52). This is called a *prothecal* fold. The apex of the long sicula is visible at the top of the photograph. Such development results in a pair of 'shoulders' before the stipes begin to grow upwards (Fig. 52), which is typical of the genus and distinguishes it from the similar *Isograptus*. Scale bar is 0.4 mm. Photo Ni Yunan.

52. The proximal details of Fig. 51 are lost in flattening of this specimen of *Pseudisograptus manubriatus* (8). Arenig, Ordovician; Macedon, Victoria, Australia; flattened specimen of form similar to that in Fig. 51: little detail can be seen. An early growth stage of the same form is seen at the bottom right. Scale bar is 1 mm.

53. Almost completely flattened, but imaged with an SEM. *Streptograptus nodifer* (4). Llandovery, Silurian; Guizhou Province, China; showing thecae so strongly retroverted that they are enrolled and the thecal apertures (not visible at all) face the dorsal side of the stipe, and were severely restricted. Scale bar is 50 μm. Photo Chen Xu, China.

54. *Paraclimacograptus pacificus* (2, 3). Ashgill, Ordovician; Kazakhstan, USSR; thecae almost orthograptid but with a slight geniculum from which grow prominent spines. The apex of the sicula is visible, at the level of the third thecal aperture of the first series, where it merges into the nema. The nema continues throughout the length of the colony, and the interthecal septa (septa between the thecal tubes) connect to it in the middle of the colony, a prominent virgellar spine projects from the proximal end and the sicular aperture bears a horizontally-directed spine (bottom right), analogous to the genicular spines along the rest of the series. Scale bar is 500 μm. Photo Tania Koren'.

55. *Rastrites* sp. (4). Llandovery, Silurian; Germany; rhabdosome was probably a helical spiral during life. Note that the distal thecae (bottom right) are drawn out into spines, which are almost certainly extended lateral tubular parts to the apertural region. Numerous siculae are scattered on the bedding plane, as well as some diminutive *Climacograptus*, small monograptid fragments, and the early growth stage of a more robust biserial graptolite (top right-hand corner). Scale bar is 2 mm. Photo Pietr Störch.

56. View of the sicular aperture and spinose thecal apertures of *Monograptus priodon* (4). Llandovery, Silurian; Cornwallis Island, Canadian Arctic; apertural view of sicula and first two thecae of the colony. The sicula has a pronounced virgellar spine and the retroverted thecal apertures (i.e. hooked to face the proximal part of the colony) bear fine laterally-directed spines. Scale bar is 50 μm.

57. Thecal cladia festoon the S-shaped stipes of *Sinodiversograptus multibrachiatus* (4). Llandovery, Silurian; Sichuan Province, China; main stipes grow in an S-shape (top of photo), one being a cladium (branch) from the sicular aperture (not visible); secondary stipes grow one from each thecal aperture, and develop similar thecae themselves. The thecae are similar to those in Fig. 53. The rhabdosomal form, S-shaped with numerous cladia, echoes earlier forms in the Ordovician, suggesting a particular hydrodynamic niche. Scale bar is 1 mm. Photo Chen Xu.

58. Cladia-bearing elegance typified by *Cyrtograptus pseudolundgreni* (4). Upper Wenlock, Silurian; Ronneburg, Germany; flattened specimen in black shale associated with numerous siculae and fragments of *Pristiograptus*. Note the strongly hooked nature of the tiny proximal thecae, the twisting of the main stipe and the two cladia. Scale bar is 2 mm. Photo Manfred Schauer.

59. Bilaterally symmetrical diplograptid. *Climacograptus medius* (2, 3). Lower Llandovery, Silurian; Weiberg, Germany; specimen with long, robust, proximal virgella and long nema; note deep thecal excavations and tectonic lineation at right angles to specimens, exaggerating the deep excavations and probably causing the slight curvature of the rhabdosome. Scale bar is 3 mm. Photo Manfred Schauer.

60. As a result of flattening the fine mesh can appear denser than it was in life. *Pseudoplegmatograptus obesus* (2, 3). Upper Llandovery, Silurian; Thuringen, Germany; showing presence of nema throughout length of colony and approximate position of thecal apertures above apertural processes (?). Note contrast with specimens in Fig. 32. Scale bar is 1 mm. Photo Manfred Schauer.

61. The beautiful *Isograptus victoriae maximus* (9). Arenig, Ordovician; Newfoundland, Canada; the nema and part of the sicula is in the process of lifting off the bedding plane surface, showing how delicately preserved some specimens can be. Ideally such a specimen should be treated with a soaking of Tragacanth gum to stick it to the rock or, alternatively, transferred to a resin block (by coating with a thick resin layer and then dissolving the rock itself in either HCl or HF as appropriate: see Ch. 11). Scale bar is 0.5 mm. Photo Henry Williams.

62. The horizontally growing, two-stiped *Expansograptus abditus* (5). Arenig, Ordovician; Newfoundland, Canada; chemically isolated specimen showing origin of first theca high on the sicula. Note the short interthecal septum of the last two thecae on the left, indicating

low overlap of the thecal tubes. Scale bar is 0.25 mm. Photo Henry Williams.

63. *Goniograptus tumidus* (5). Bendigonian (Arenig), Ordovician; Campbelltown, Victoria, Australia; specimen showing thecal profile well, and the difficulty of seeing thecae proximally, and hence development there (bottom left) other than the combination of dichotomies and lateral branching. Scale bar is 3 mm.

64. Tangled mat of pendent *Tetragraptus fruticosus*, associated with the very slender *Sigmagraptus* (5). Arenig, Ordovician; Bendigo, Australia; sicula can be identified at the proximal end of several rhabdosomes, and the outlines of the thecae, but not the growth details. Miners in the old gold mining town of Bendigo used the rhabdosome patterns to work out a sequence of rocks in the shafts, and hence deduce the structure of the mine, despite having no access to scientific literature describing graptolites. Scale bar is 3 mm.

65. *Pseudoplegmatograptus obesus* (4). Llandovery, Silurian; Canadian Arctic; showing prosicula with longitudinal strengthening rods, an entire periderm (compare same specimen in Fig. 32 and same species in Fig. 60), and possibly the prosicula/metasicula junction (near bottom of page). Scale bar is 25 μm. Photo Michael Melchin.

66. *Paraplectograptus macilentus* (4). Wenlock, Silurian; Canadian Arctic; simple clathrial network (contrast Fig. 67) with minimal indication of thecae. Nema and ancora present. Note granular/pustulose ornament of lists. Scale bar is 150 μm. Photo Michael Melchin.

67. Small, dense network nonetheless shows traces of thecal outlines in *Gothograptus eisenacki* (4). Wenlock, Silurian; Canadian Arctic; proximal end is at the foot of the page and the (complete) colony tapers to a thin meshwork tube distally, attached to the nema. This tube is called an *appendix*. Proximally the nema divides at its extremity to a series of lists, called the *ancora*, and above it the clathrial network of lists roughly define the thecal outlines and walls. Much more reduction of the periderm could, of course, give rise to entirely soft-bodied graptolites. Scale bar is 150 μm. Photo Michael Melchin.

68. The scaffold in detail. *Pseudoplegmatograptus obesus* (4). Llandovery, Silurian; Canadian Arctic; enlargement of lists showing the seams referred to in Ch. 3 from which the thin periderm layer has fallen away. Scale bar is 25 μm. Photo Michael Melchin.

69. Inside the sieve, looking down the colony of *Retiolites geinitzianus densireticulatus* (4). Llandovery, Silurian; Canadian Arctic; proximal end of specimen shown in Fig. 69; approximate positions of thecae are shown, and also the tendency for the lateral rhabdosomal walls to develop a larger pore at intervals (seen to perfection in Fig. 71). Scale bar is 200 μm. Photo Michael Melchin.

70. Sieve-like pores perforate the thecal walls in *Retiolites geinitzianus densireticulatus* (4). Llandovery, Silurian; Canadian Arctic; distal view of colony showing internal arrangement of lists, giving an overall impression of great strength, possibly explaining why such retiolitids are cosmopolitan and occur in a wide range of depositional environments. Scale bar is 100 μm. Photo Michael Melchin.

71. Enlarged lateral pores typify *Stomatograptus grandis* (4). Llandovery, Silurian; Canadian Arctic; showing lateral pores characteristic of the genus, thecal outlines, and thickened apertural apertures. Scale bar is 200 μm. Photo Michael Melchin.

72. What everyone thinks of as a graptolite; *Monograptus marri* (4). Llandovery, Silurian; Ménez-Bélair, France; latex of external mould of pyritized specimen. Note preparation marks made by razor blade, near proximal end, the presence of the sicula, and the slightly different appearance of the larger of the two flattened specimens to the right of the main specimen. Scale bar is 1 mm. Photo Florentin Paris.

73. Named after one of the editors! *Monograptus rickardsi* (4). Llandovery, Silurian; Ménez-Bélair, France; pyrite internal mould showing thecal overlap, interthecal septum, and common canal clearly, with fusellar growth increments visible in places. The species has apertural spines but these are not preserved on the internal mould. Scale bar is 1 mm. Photo Florentin Paris.

74. One of the many kinds of fish hook shaped rhabdosomes, usually in the Llandovery, in this case with the thecae on the outside of the curve. *Monograptus triangulatus* (hooked rhabdosome) and *Coronograptus gregarius* (4). Low Llandovery, Silurian; central Wales; pyritised specimens showing partial infill of thecae. Note also the tectonic lineation, along the length of the page, the compression related to which has disrupted the specimens. Scale bar is 1 mm.

75. Thecae on the inside of the curve. *Monograptus plumosus* (4). Llandovery, Silurian; central Wales; pyritised specimen showing granular or microcrystalline pyrite, prothecal folds on the specimen and metathecal enrolling. Note the incomplete preservation of the proximal region. Scale bar is 1 mm.

76. Chemically isolated specimens (contrast Fig. 75) reveal detail not otherwise visible. *Monograptus exiguus* (4). Llandovery, Silurian; Sweden; showing paired and bifurcating thecal processes, rarely seen on specimens not isolated from the rock. Note also prothecal folds, each grooved down the middle (dorsal edge) because they have grown around a preformed, straight nema (seen at the distal extremity, top of page), just as growth of the prothecal fold is beginning. The virgella is broken off the sicula. Scale bar is 250 μm.

77. A swarm of the incurved *Testograptus testis* (4). Wenlock, Silurian; Ronneburg, Germany; flattened, spinose specimens, ventrally curved, preserved in black shale with an abundance of secondary pyrite crystals (black specks). Scale bar is 4 mm. Photo Manfred Schauer.

78. The astogeny here exhibits marked changes of thecal form. *Monograptus planus* (4). Llandovery, Silurian; Ronneburg, Germany; numerous proximal ends, early growth stages in a dark siltstone; the changing shape of the thecae from slim proximal end to robust distal parts is apparent. Tectonic deformation lineation runs from top left to bottom right, and dimensional changes relative to this can be seen. Scale bar is 2 mm. Photo Manfred Schauer.

79. The characteristic watch spring of *Monograptus spiralis* (4). Llandovery, Silurian; Ronneburg, Germany; cosmopolitan uppermost Llandovery species in which the rhabdosome is arranged in either a plane spiral or a low helical spiral. Thecae are very distinctively triangular and spinose. Isolated specimens show the spines to be paired and directed laterally, not ventrally as seen here (which is a result of flattening). Scale bar is 2 mm. Photo Manfred Schauer.

80. One of the last graptolites on earth, *Pseudodictyonema* sp. (Dendroidea of Text figure 1 and appendices 3 and 4). Upper Carboniferous; Pendle Hill, England; robust and regular dissepiments can be seen, but not thecal details on this specimen. Scale bar is 5 mm.

81. An efficiency in feeding strategy underlies the subtle geometry of stipe dichotomy. *Loganograptus rectus* (5). Bendigonian (Arenig), Ordovician; Campbelltown, Victoria, Australia; rare specimen of horizontally disposed dichograptid with sicula preserved. Note that the first two stipes diverge at 180° from the tiny sicula before the first dichotomy. The distance/region *between* the first two dichotomies is called the *funicle*. Scale bar is 3 mm.

82. *Clonograptus persistens* (5). Bendigonian (Arenig), Ordovician; Campbelltown, Victoria, Australia; classic clonograptid dichotomies with short funicle, four conspicuous funicular areas and dichotomies of decreasing frequency distally on any one line of development, thus decreasing the chances of distal overgrowth of stipes caused by too frequent branching (see Ch. 12). Scale bar is 5 mm.

83, 84. Thickening of the proximal regions gives added strength to the growing, multibranched colony of *Clonograptus trochograptoides* (Anisograptidae in Text figure 1 and appendices 3 and 4); Bendigonian (Arenig), Ordovician; Campbelltown, Victoria, Australia; regular dichotomies of the stipes are the main feature; typical fret-saw appearance of thecae only visible at periphery of colony (84), where the stipes twist over and lie on their sides in full profile; much secondary, cortical thickening has taken place in the proximal (central) region of the colony. Some of these early planktonic forms reached a diameter of 50 cm or more. Scale bar is 2 mm.

85. Contrasting forms of contemporaneous genera. *Pseudotrigonograptus ensiformis* (10) associated with *Isograptus* (top right) (9) and *Pseudisograptus* (top centre) (8) and extensiform didymograptids (5). Upper Arenig, Ordovician; Newfoundland, Canada; note that the specimens are flattened with a tendency for the carbonised periderm to flake off (see remarks in explanation of Fig. 61). Scale bar is 2 mm.

86. *Adelograptus victoriae* (Anisograptidae of Text figure 1 and appendices 3 and 4) associated with *Tetragraptus decipiens* (top left corner and bottom right corner) (5). Lancefieldian (Tremadoc), Ordovician; Lancefield, Victoria, Australia; some of the earliest known chloritic preservations in black shale. Scale bar is 3 mm.

87. Compact and linear colony forms. *Phyllograptus anna* (Phyllograptidae in appendices 3 and 4). Bendigonian (Arenig), Ordovician; two specimens show the virgellar spine proximally which distinguishes *Phyllograptus* from *Pseudophyllograptus*; associated with three-stiped *Tetragraptus fruticosus* (to right), *T. pendens* (slim form near *T. fruticosus* specimen), *Sigmagraptus* (bottom right near *Phyllograptus* specimen), and declined *Didymograptus* or *Acrograptus*. Scale bar is 1 mm.

88. Reflexed and pendent colony form. *Tetragraptus serra* (main specimen) (5) associated with *T. fruticosus* (top of page) and *Expansograptus* (cutting across *T. serra*); Bendigonian (Arenig), Ordovician; Bendigo, Victoria, Australia; reclined tetragraptid with four robust stipes diverging upwards from prominent sicula. Scale bar is 1 mm.

89. Three branches pendent. *Tetragraptus fruticosus* (5). Bendigonian (Arenig), Ordovician; Bendigo, Australia; note that the middle stipe shows no thecae: this is because

it is a dorsal view of the stipe and the thecae face into the rock, in contrast to the other two stipes which are seen in full profile. Scale bar is 1 mm.

90. Three branches reclined. *Tetragraptus phyllograptoides* (5). Arenig, Ordovician; Vastergotland, Sweden; reclined, three-stiped specimen with thecal details obscure on middle stipe due to dorsal view presented (see Fig. 89 also). Scale bar is 1 mm. Photo Jörg Maletz.

91. *Tetragraptus approximatus* (5). Arenig, Ordovician; Newfoundland, Canada; although the sicula is missing on this H-shaped species, it is now known that the four stipes have a horizontal disposition and that the sicula, therefore, comes out of the page (or into it). Scale bar is 1 mm. Photo Henry Williams.

92. Possibly another experiment in weight reduction. *Kinnegraptus kinnekullensis* (5). Arenig, Ordovician; Newfoundland, Canada; very slender prothecae are typical, as is the expanded apertural region of thin periderm; specimen has a conspicuous nema. Note the zig-zag fusellae on the ventral apertural processes. Chemically isolated specimen, light microscope photography. Scale bar is 250 μm. Photo Henry Williams.

93. *Expansograptus constrictus* (5). Arenig, Ordovician; Vastergotland, Sweden; note that the sicula has pulled away (and is on the counterpart slab); and that the thecal tubes widen where they have been flattened, near the apertures. Species is gently reclined. Scale bar is 1 mm. Photo Jörg Maletz.

94. Neat proximal development seen in *Didymograptus vacillans* (5). Arenig, Ordovician; Vastergotland, Sweden; latex cast of three-dimensional specimens. Note the reverse views of the specimens in the top half of the page, showing the early thecal development and crossing canals quite clearly; and the obverse views at the bottom of the page showing the sicula in full view. Scale bar is 1 mm. Photo Jörg Maletz.

95. Uniserial, fully scandent *Monograptus nimius* (4). Pridoli, Silurian; Kazakhstan, USSR; biform colony, uniserial, scandent, with long, simple, tubular thecae through much of the colony, but rounded thecal apertures for the first two thecae. Note the long interthecal septae, the black nemal rod on the dorsal (anti-apertural) side, and the common canal joining the bases of all thecal tubes (it is situated adjacent and parallel to the nemal rod). Scale bar is 1 mm. Photo Tania Koren'.

96. *Expansograptus protobalticus* (5). Arenig, Ordovician; Vastergotland, Sweden; note presence of short nema and early origin of first theca high on the sicula. Scale bar is 2 mm. Photo Jörg Maletz.

97. Two branches fully scandent are still detectable in the network of *Pseudoplegmatograptus obesus* (4). Llandovery, Silurian; Canadian Arctic; full profile of proximal end showing prosicula, nema, and broad thecal outline and apertural regions. Scale bar is 250 μm. Photo Michael Melchin.

98. Bottom current orientation of *Climacograptus extraordinarius* (2, 3). Latest Ordovician; Kazakhstan, USSR; the six current oriented specimens all have their proximal ends pointing to the bottom of the page, indicating a current from that direction. This species is one of the most cosmopolitan and short-lived species known, and is vital in world-wide correlation of the Ordovician-Silurian boundary. It occurs also at Dob's Linn (see Appendix 2). Scale bar is 1 mm. Photo Tania Koren'.

99. *Diplograptus magnus* (2, 3). Low Llandovery, Silurian; central Wales; three dimensional pyritised specimen; tectonic lineation left to right, resulting in a specimen folded/crumpled at right angles to its length; obverse view. Scale bar is 2 mm.

100. *Rhaphidograptus toernquisti* (2, 3). Low and Middle Llandovery, Silurian; Prague, Czechoslovakia; perhaps the most common and cosmopolitan of all Silurian graptoloids. Note the virgella proximally, the exposed sicula (to left, at proximal extremity) and the absence of a theca (bottom left of specimen): this is not an artifact, and the genus is typified by suppression of the second thecae of the rhabdosome. Scale bar is 1 mm. Photo Pietr Störch.

101. Oblique view of thecal apertures in well preserved *Climacograptus normalis* (2, 3). Low Llandovery, Silurian; Morocco; specimen in low relief, sub-apertural view of first thecal series; excavations and geniculum visible, as is median septum on this reverse view; sicula is seen faintly 'pressed through'; thecal overlap is approximately one half. This is typical of a number of common and cosmopolitan, diminutive, non-spinose, low Silurian climacograptids. Scale bar is 250 μm.

102. *Cephalograptus cometa* (2, 3). Middle Llandovery, Silurian; Ronneburg, Germany; note the extremely elongate thecal tubes which open distally; only ten thecae comprise the full colony; the sicula is tiny and is obscured in this specimen. In life it seems likely that the thecae were pendent from the sicular region, rather than above it as depicted here. Scale bar is 1 mm. Photo Manfred Schauer.

103. *Pristiograptus nudus* (4). Upper Llandovery, Silurian; Ménez-Bélair, France; mould of simple thecal tubes (which persisted right through the Silurian); latex of external mould. Scale bar is 1 mm. Photo Florentin Paris.

104. *Pseudoclimacograptus* sp. (2, 3). Upper Llandovery, Silurian; Ménez-Bélair, France; an undescribed pseudoclimacograptid with striking zig-zag median septum and accompanying undulating growth of thecal tubes; reminiscent of Ordovician forms rather than later forms of the genus. Scale bar is 500 μm. Photo Florentin Paris.

105. Delayed dichotomies, at times, prevents overcrowding. *Anisograptus matanensis* (Anisograptidae of Text figure 1 and appendices 3 and 4). Lower Tremadoc, Ordovician; Newfoundland, Canada. Scale bar is 1 mm. Photo B-D. Erdtmann.

106. Possibly a teratological specimen of *Petalograptus ovatoscopularus* (2, 3). Llandovery, Silurian; Thuringen, Germany; note the excessively divided nema and the apparent divergence of the later part of the two scandent stipes; morphology and function of this species are relatively unknown. Scale bar is 2 mm. Photo Manfred Schauer.

107. Remarkably prolonged nemata occur in some species, seen here in *Petalograptus ovatoelongatus* (2, 3). Llandovery, Silurian; Thuringen, Germany; nema also has a distal thickening. Scale bar is 2 mm. Photo Manfred Schauer.

108. *Dimorphograptus confertus* (4). Llandovery, Silurian; Vseradice, Czechoslovakia; note uniserial/biserial stipe, numerous siculae, monograptids and climacograptids on the slab. Scale bar is 1 mm. Photo Pietr Störch.

109. Tangled and enmeshed *Petalograptus fusiformis* (2, 3). Llandovery, Silurian; Sichuan Province, China; this assemblage almost certainly represents a collapsed synrhabdosome, an association during life of a number of colonies (see also Fig. 110). Note the pronounced nemata, the tangle of them near the centre of the photo, and the non-profile (or apertural) view of the dark specimen, also near the centre of the cluster. Scale bar is 5 mm. Photo Chen Xu.

110. A clear biological association of mature colonies of the same species. *Saetograptus varians* (4). Ludlow, Silurian; central Wales; synrhabdosome with tangled nemata at the centre. Synrhabdosomes are almost always monospecific and with similar growth stages involved, as here. Scale bar is 1 mm.

111. The unmistakable Y-form of *Oncograptus upsilon* (5). Arenig, Ordovician; Gisborne, Victoria, Australia; classic Y-shaped rhabdosome, lacking a nema, almost certainly lived 'upside down', that is bell-like from the sicula. Scale bar is 2 mm.

112. One of the largest and rarest dichograptids known. *Orthodichograptus robbinsi* (5). Ordovician; Bendigo, Victoria, Australia; but it occurs in both the northern and southern hemispheres even though only 30 specimens have ever been collected. Near the origin is a discontinuous web structure (bottom right), which would certainly have retarded sinking of such a huge colony. Fret-saw appearance of thecae in profile is visible in places and, inconspicuous in the background are specimens of *Tetragraptus*, *Loganograptus*, *Phyllograptus* and *Sigmagraptus*. The dark crescentic areas are remnants of dark colouring matter of the original black shale, the remainder having been bleached out along the stipes themselves. Scale bar is 5 mm.

113. Striking wing-like processes typify some graptolites, as in *Climacograptus longispinus* (2, 3). Ashgill, Ordovician; Kazakhstan, USSR; climacograptid thecal excavations clearly visible along the sides of the colony which is, however, dominated by a spectacular pair of proximal vanes. Recently hydrodynamic modelling studies suggest that such structures were involved with feeding rotation of the rhabdosome. Scale bar is 400 μm. Photo Tania Koren'.

114. *Plegmatograptus nebula latus* (4). Ashgill, Ordovician; Kazakhstan, USSR; 'retiolitid' in which the detailed structure is not yet understood; note bifurcating apertural processes, external lacinial network, and nema preserved faintly. Scale bar is 1 mm. Photo Tania Koren'.

115. Identical twins: *Orthograptus amplexicaulis* (2, 3). Ashgill, Ordovician; Kazakhstan, USSR; classic orthograptid with simple tubular thecae, proximal spines on $th1^1$ and sicular aperture, plus a pronounced virgella. Interthecal septae and common canal adjacent to nema well seen. Scale bar is 1 mm. Photo Tania Koren'.

116. Diagenetic bleaching of the matrix surrounding *Loganograptus logani* (5). Bendigonian (Arenig), Ordovician; Castlemain, Victoria, Australia; note the bleaching of the black shale along the positions occupied by the stipes; this is a half way stage to that seen in Fig. 112. Scale bar is 5 mm.

117. The zig-zag stipes of *Goniograptus thureaui* (5). Bendigonian (Arenig), Ordovician; Campbelltown, Victoria, Australia; associated with two species of tetragraptid and a

Pseudobryograptus (top right). Note how the other graptolites at the bottom of the page overlay the *Goniograptus* on a slightly different bedding plane. Scale bar is 3 mm.

118. A flattened jumble of conical rhabdosomes of *Monograptus turriculatus* (4). Llandovery, Silurian; Howgill Fells, England; slab of variously preserved and deformed specimens of a helically spiralled graptolite. Note that some cones are preserved in plan view and others in profile. Scale bar is 3 mm.

119. The lop-sided *Azygograptus* sp. (5). Llanvirn, Ordovician; Lake District, Cumbria, England; the ultimate in dichograptid stipe reduction with a single gently declined stipe. Note the short nema. Scale bar is 200 μm.

120. Contrasting planktonic form, different life habits, yet nevertheless preserved together. *Sigmagraptus crinitus* (7). Arenig, Ordovician; main stipes with long lateral branches, one developed from each theca, associated with *Phyllograptus* sp. and pendent *Tetragraptus* fragments. Scale bar is 5 mm.

121. The earliest strongly reclined graptolite. *Psigraptus jacksoni* (Anisograptidae of Text figure 1 and appendices 3 and 4). Tremadoc, Ordovician; Tasmania, Australia; first few thecae have strongly isolate apertural regions, as has the sicula (centre of photo.). This specimen is from the same locality as that which yielded the *Psigraptus* zooids (Fig. 34). Scale bar is 500 μm.

122. A carbonate rock almost solid with graptolites. *Rhabdinopora* sp. (Anisograptidae of Text-figure 1 and appendices 3 and 4). Tremadoc, Ordovician; Newfoundland, Canada; mats of specimens on successive bedding planes in a fine grained limestone; main specimen, filling most of field of view, has a sicula preserved at its apex. Scale bar is 5 mm.

123. Graptolitic shale *par excellence*, with abundant specimens on several bedding planes, with dark, pyritous nodules, and many specimens seen edge on in the dark area towards the bottom right (4). Wenlock, Silurian; France; two main thecal types are visible: excavations of *Monoclimacis* type and hooks of *Monograptus* type. A strong tectonic deformation is present, elongating specimens parallel to the length of the page, and flattening them at right angles to this direction. Scale bar is 5 mm.

124. *Phyllograptus* sp. (5). Arenig, Ordovician; Newfoundland, Canada; specimens preserved in three dimensions, with original periderm intact, in a crystalline limestone. Scale bar is 2.5 mm.

125. Stranded amongst shells. *Climacograptus* sp. (2, 3). Llandovery, Silurian; Anticosti Island, Canada; somewhat battered specimen preserved in shelly limestone deposited in an inshore shelf environment. Scale bar is 1 mm.

126. Damage limitation. *Rhabdinopora flabelliformis anglica* (Dendroidea of Text figure 1 and appendices 3 and 4). Tremadoc, Ordovician; Shropshire, England; specimen regenerated, after damage during life, to the distal region. Scale bar is 3 mm.

127. *Monograptus runcinatus* (4). Llandovery, Silurian; Ménez-Bélair, France; specimen shows slight prothecal folding and a tightly enrolled hook, with no thecal overlap; sicular well seen on a fragment near the bottom of the photograph; latex preparation. Scale bar is 400 μm. Photo. Florentin Paris.

128. Preparation can uncover all. *Rhaphidograptus toernquisti* (2, 3). Llandovery, Silurian; central Wales; note absence of second theca of rhabdosome, presence of sicula, delayed medium septum which comes in opposite the ninth theca of the first thecal series, and the preparatory work that has been done with blade and needle to expose the specimen fully. There is a tectonic lineation running across the page and the deformation has caused some slight crumpling of the specimen. Scale bar is 1 mm.

129. In this specimen matrix still obscures part. *Monograptus marri* (4). Llandovery, Silurian; Ménez-Bélair, France; notice pyrites *external* to graptolite in places; *M. marri* is typified by thecae whose apertures face proximally, with little apertural narrowing, in where the hook of the theca is like an inverted U. Scale bar is 1 mm. Photo Florentin Paris.

130. *Monograptus parapriodon* (4). Llandovery, Silurian; Ménez-Bêlair, France; a short length of thecal overlap, increasing in distal thecae and a more beak-like thecal hook than in *M. marri* (Fig. 129) emphasizes the 'slight' specific distinctions that are possible – and which are of great use in stratigraphy. Scale bar is 1 mm. Photo Florentin Paris.

131. Mounted in a glass slide for posterity, such a specimen could last for ever. *Climacograptus brevis* (2, 3). Caradoc, Ordovician; Scotland; chemically isolated and cleared (Ch. 11) but delicately preserved specimen showing prosicula, metasicula (with closely spaced fuselli), virgella, nema, zig-zag fusellae on thecae (ventral walls) and faint but numerous cortical bandages. Scale bar is 115 μm.

132. Ultra thin slices of periderm at enormous magnification produce surprising results. *Dictyonema rarum* (Dendroidea of Text figure 1 and appendices 3 and 4). Ordovician; Sweden; TEM photographs, X20,000 showing collagen fibrils in the cortical bandages (each major boundary is the edge of a bandage). The small cross bars connecting the fibrils are actually sections of the annulations which ring each fibril. The more spongy textures seen in places is the texture of collagen seen in the fuselli: it both mixes in with cortical fibrils, as here, and grades into it (see Ch. 3 for full significance). Scale bar is 0.5 μm above and 0.8 μm below. (Published with permission of the Palaeontological Association).

133. The possible life positions of monograptids. *Monograptus* sp. (4). Llandovery, Silurian; Dalarne, Sweden; specimens in process of chemical isolation from limestone fragments, by means of HCl. Note the oily scum at the top of the flask, the fragment of fretted limestone at the bottom, and the almost vertical orientation of the stipes. It is likely that straight monograptids so oriented in life position. Scale bar is 5 mm and 10 mm.

134. A tiny fragment of life 'frozen' in perpetuity. *Graptoblastoides* sp. (Crustoidea in Text figure 1 and appendix 3). Ordovician; central Poland; normal fusella fabric is seen to the left, with spongy texture typical of fuselli, and these layers enclose (to the right) partly degenerate cell tissues, perhaps including yolk bodies as well as cell membranes. The denser material in the top right and bottom right corners further bounds the cell tissue within the thecae; its nature is unresolved at present. Scale bar is 4 μm. (Published with the permission of Cambridge University Press).

135. The colony's vital fluids are shared through these microscopic tubes. *Acanthograptus musciformis* (Dendroidea of Text figure 1 and appendices 3 and 4). Upper Ordovician; Ojle Myr, Gotland, Sweden; section across a stolon system just above the *node* referred

to in the text (Ch. 1). The large diameter, thick-walled tube at the bottom of the page is the *autothecal stolon*; just above it is a thinner walled tube, almost triangular in shape, which is the *stolothecal* (or parent) stolon; top left is a smaller hole surrounded by a mass of pale coloured peridermal material, which is the *bithecal* stolon as seen exactly at the point where it blossoms into a bithecal tube (i.e. this photograph is of the very base of that tube where it rounds off and joins the hollow stolon). One mm or so above this level the autothecal stolon would also give rise to an autothecal tube sitting on the stolon; and a mm or so above that the stolothecal stolon would again divide into three, at a nodal thickening. Scale bar is 2 μm.

136. A graptolite as clastic debris. *Monograptus sedgwickii* (4). Llandovery, Silurian; County Down, Northern Ireland; we interpret this striking preservation in a different way to Jeremy Smith's original idea (Geological Magazine 1957, 425–428) which was that the graptolite was pushed along the bottom, scratching the grooves with its thecae. In fact the grooves line up with the (topographical) low points *between* the hooks, which would in any case have been anchored by the lateral spines, and the ridges of silt are in the shadow of the hooks, exactly where sediment would accumulated. Occurrences of three dimensional graptolites in sand are quite varied in preservation, and this is only one of those seen (see also Fig. 40). Scale bar is 2 mm. (Published with the permission of the editors of Geological Magazine).

137. The classic Arenig/Llanvirn locality of Abereiddi Bay in Pembrokeshire. Access is via a track and small stream just off the photo to the right; the bedding planes to the left, and their continuations in the foreshore, yield abundant tuning fork graptolites (see Appendix 2).

138. Classic, laminated graptolitic mudstone, bioturbated towards the top by *Chondrites* tubes; strata totally barren of graptolites is at the top right corner; also bioturbated, these beds are more aerobic at deposition and yield a benthos; the laminated beds contain primary pyrite, secondary pyrite (cubes) and a percentage weight of free carbon of about 5%. The age of this particular slab is basal Wenlock, and the locality the Howgill Fells, Cumbria, England. Scale bar is 5 mm. (Published with the permission of the editors of Geological Magazine).

PLATES

1

2

3

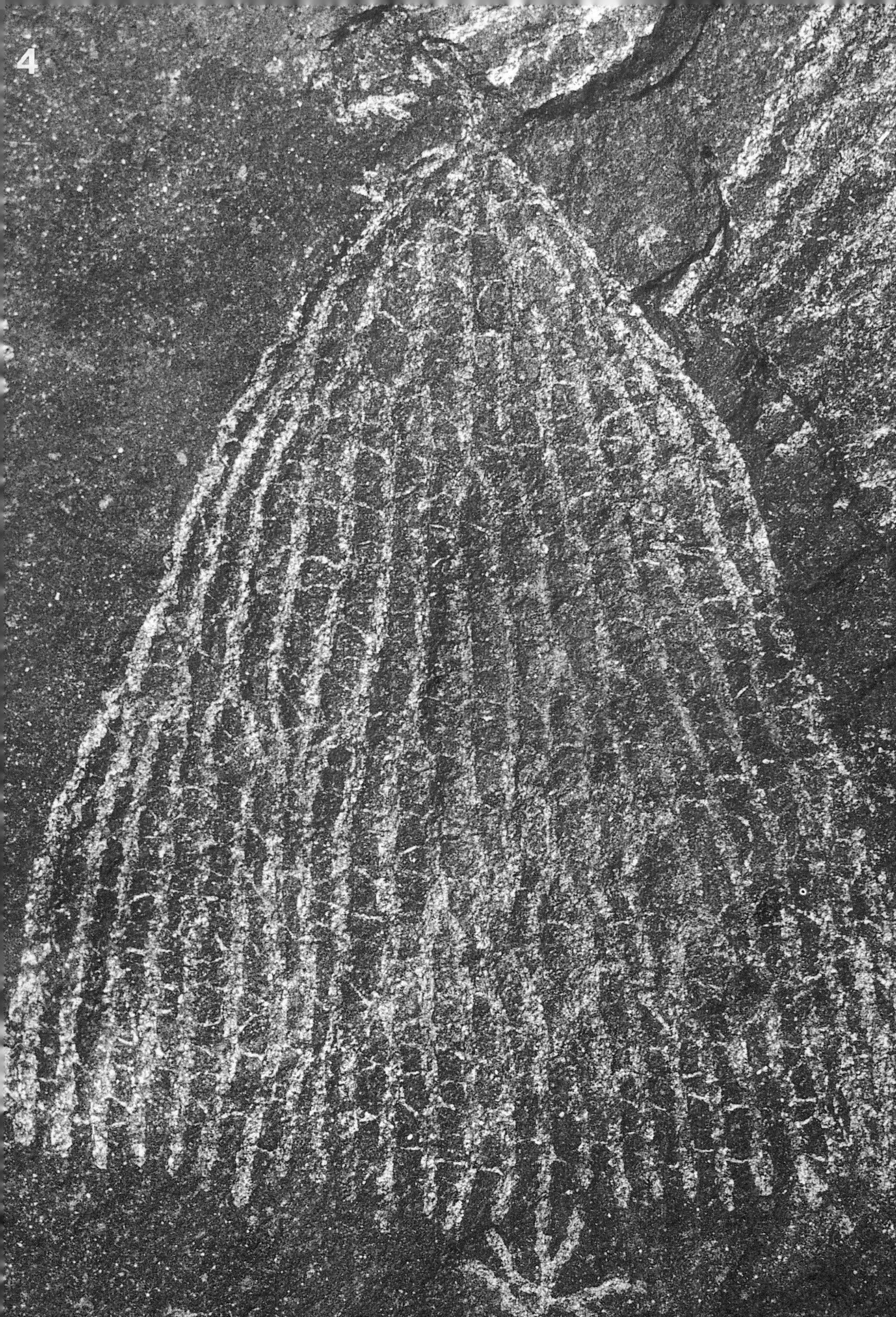
4

5

6

7

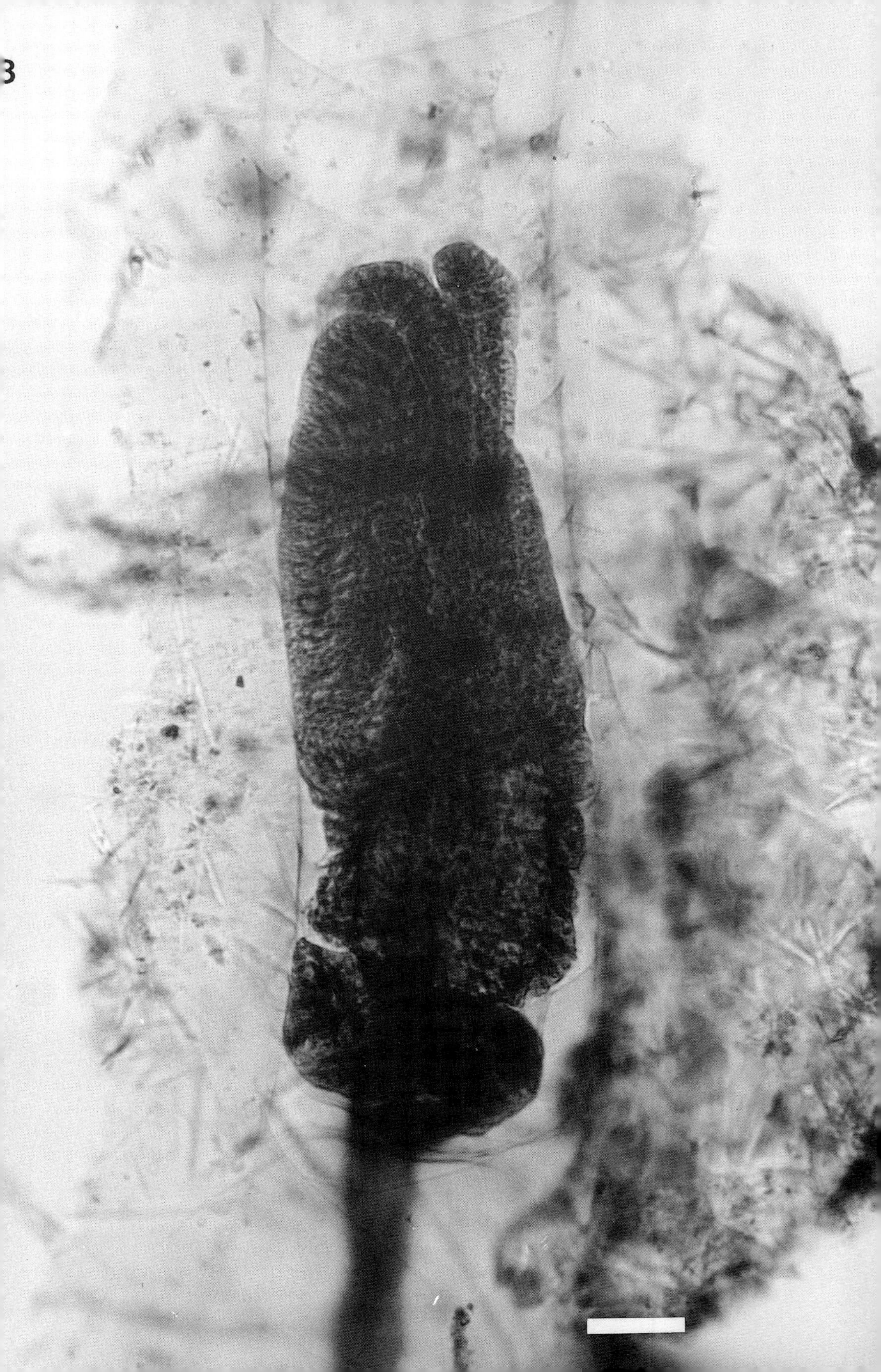

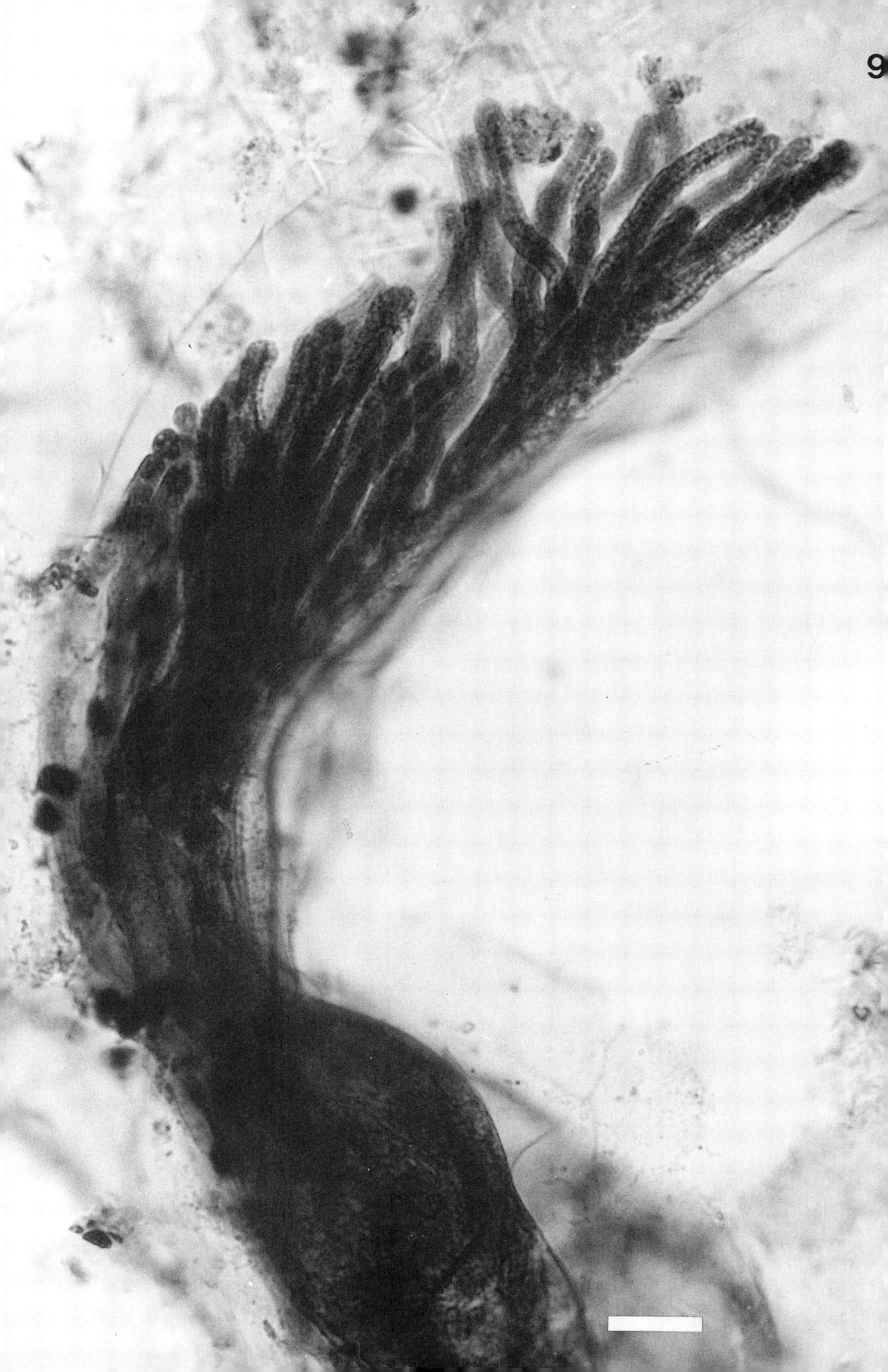
9

10

11

12

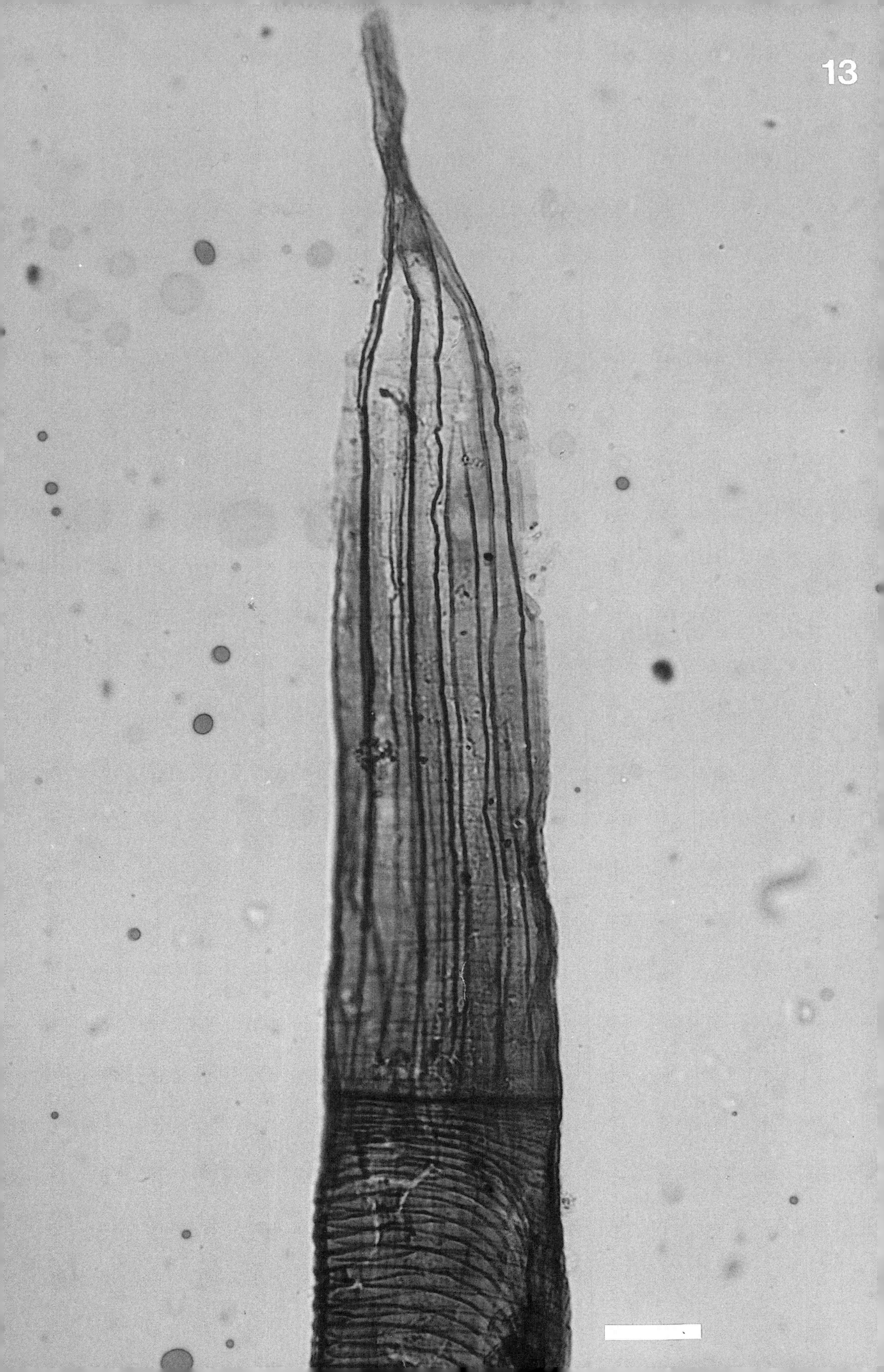
13

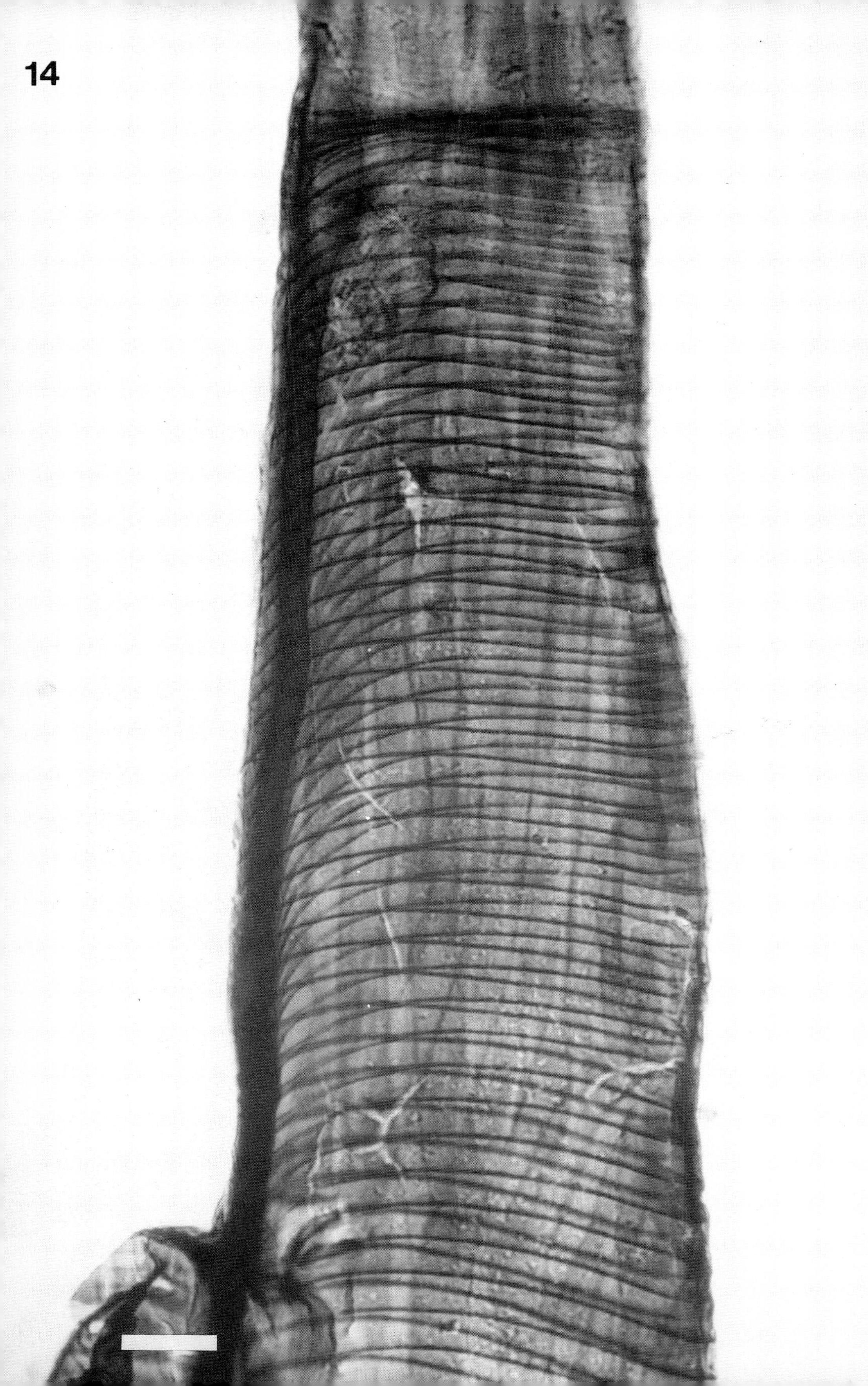
14

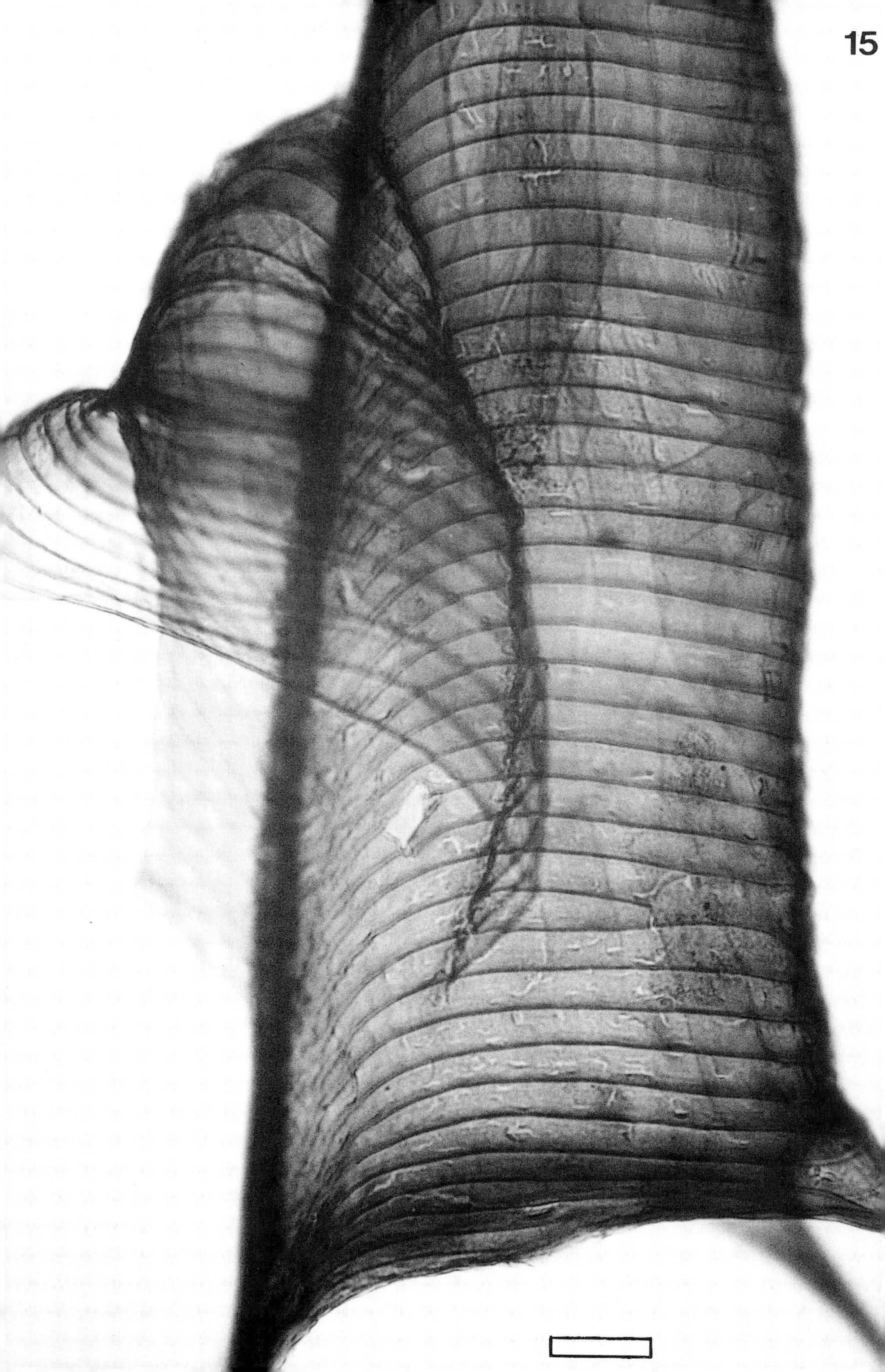
15

16

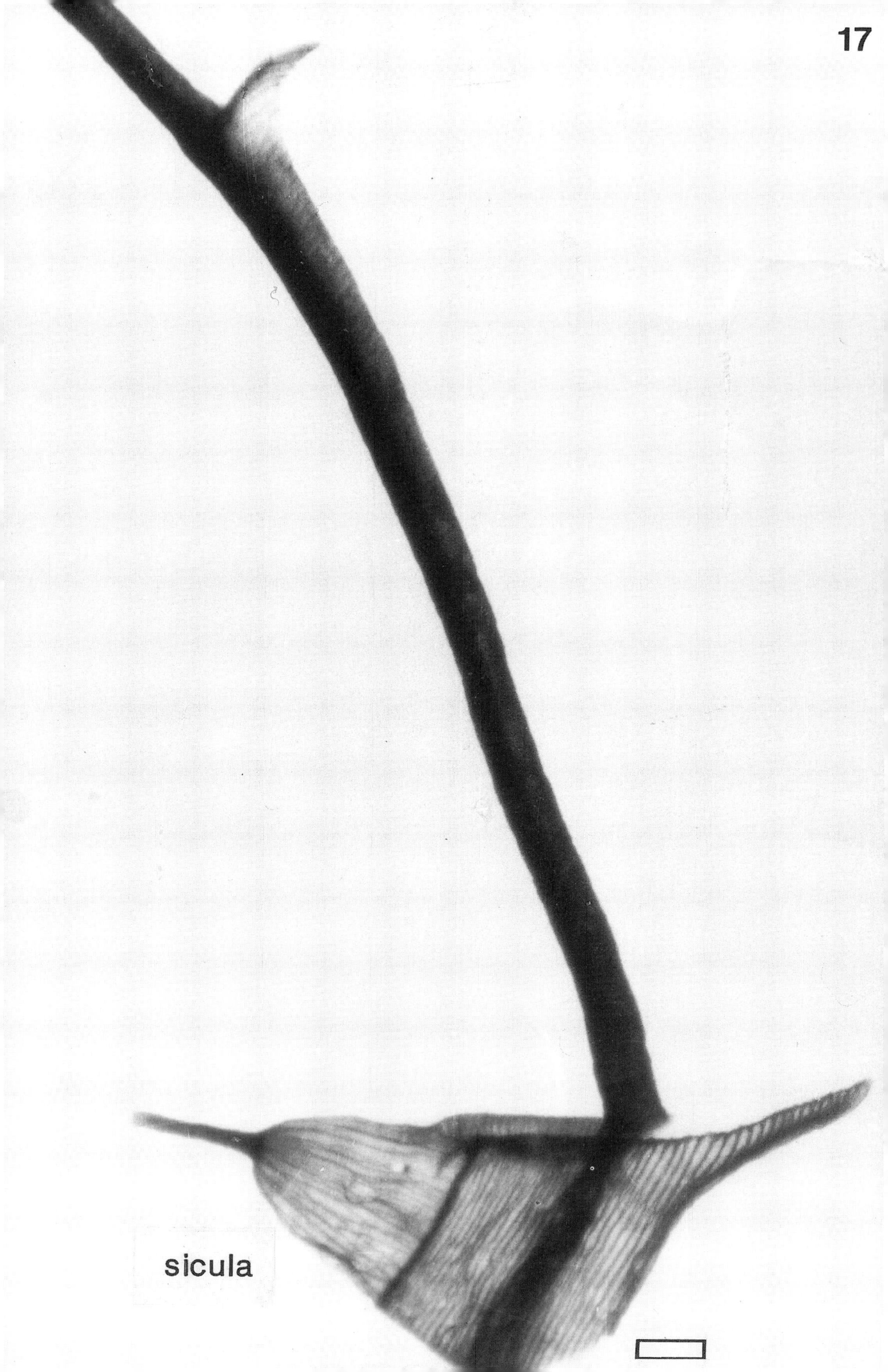
17
sicula

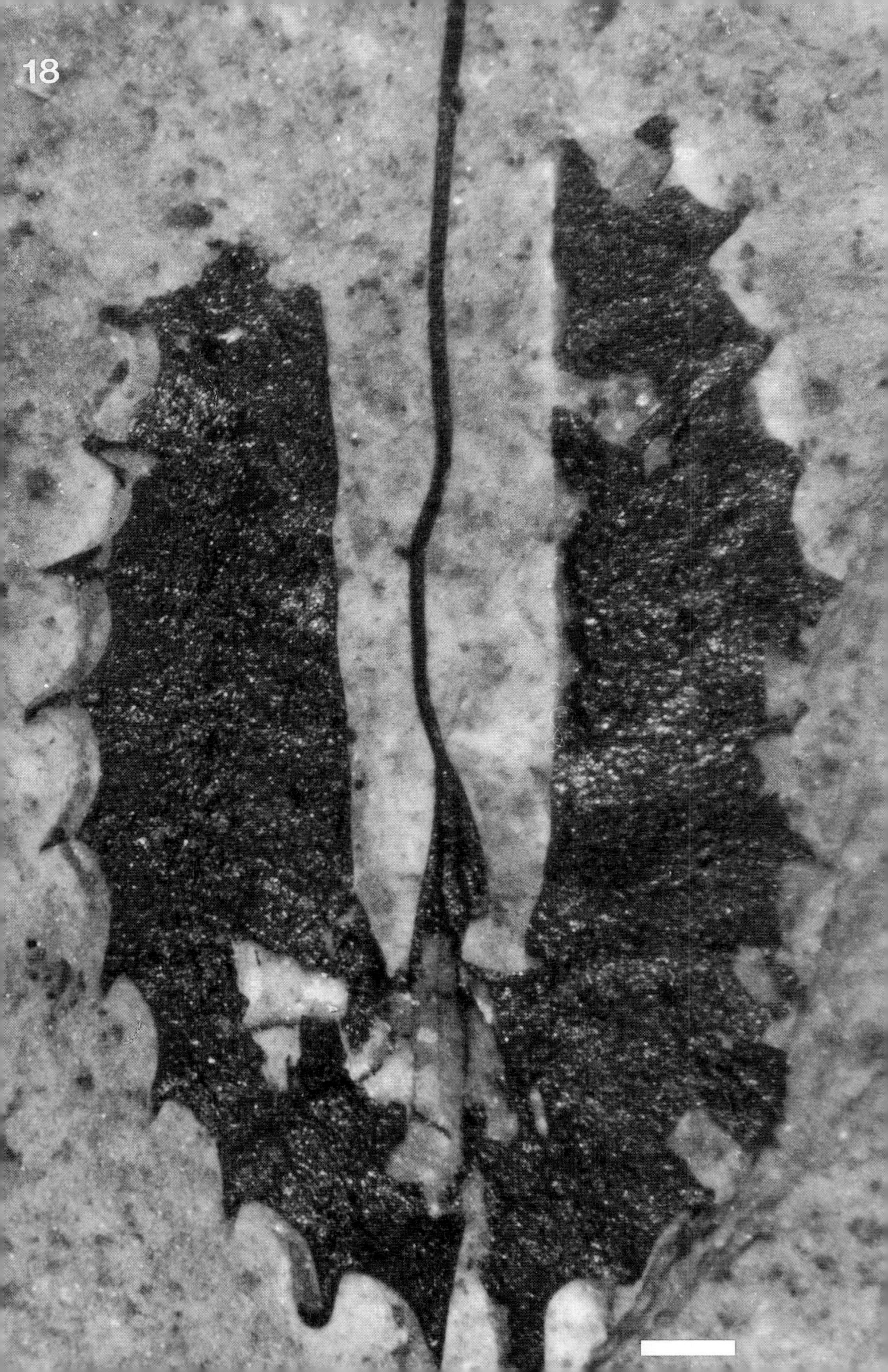
18

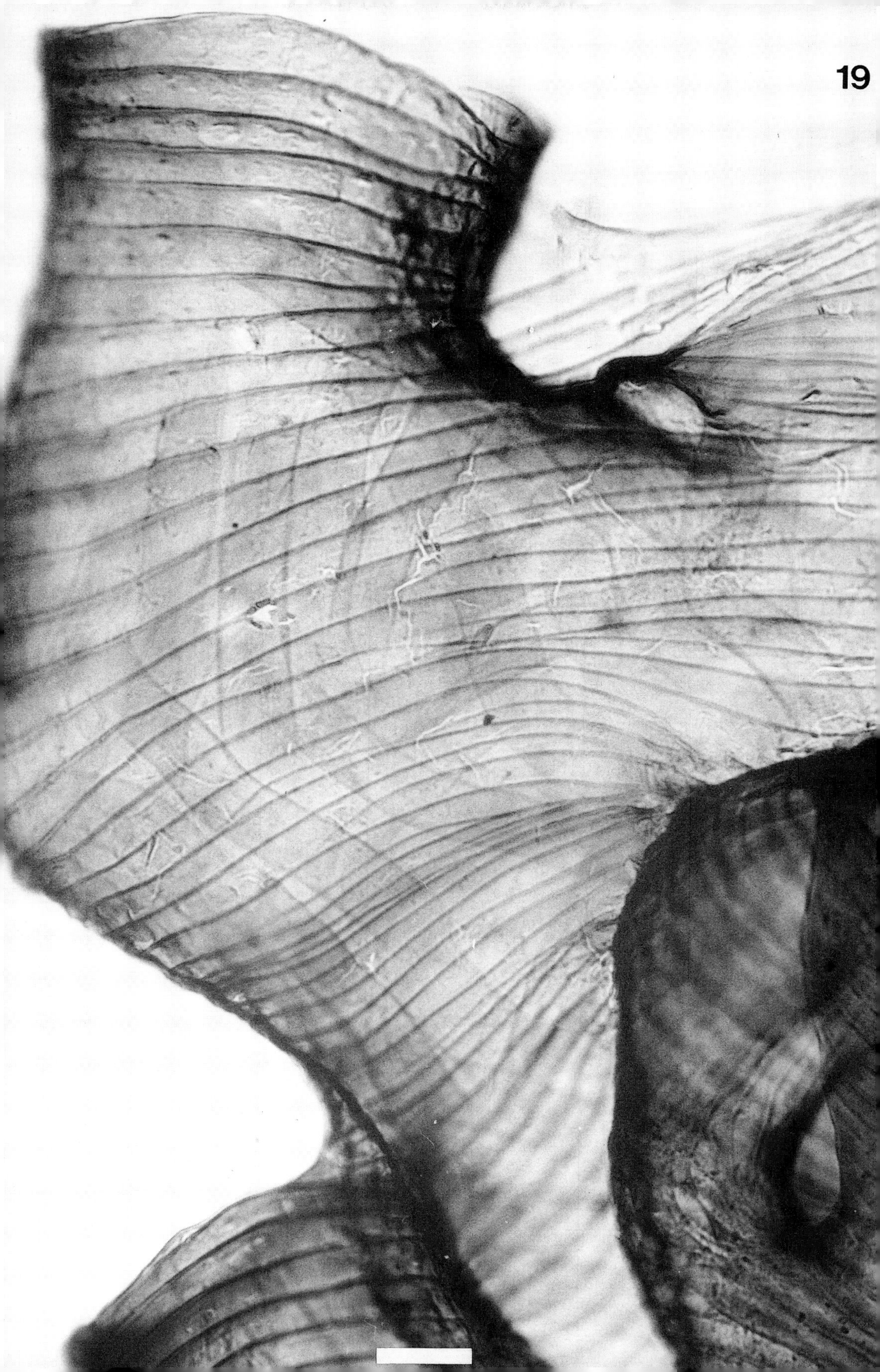
19

20

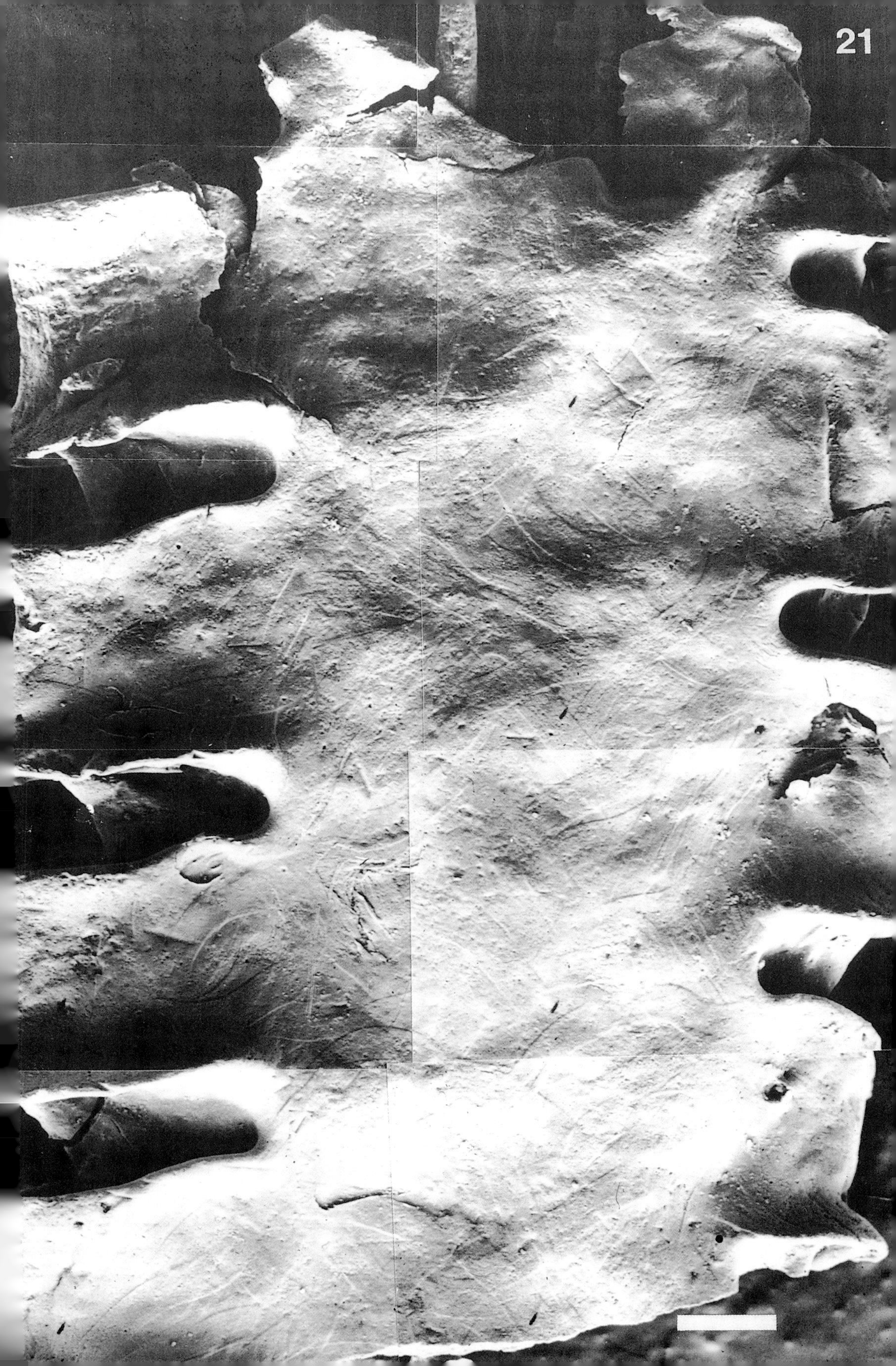
21

22

23
sicula

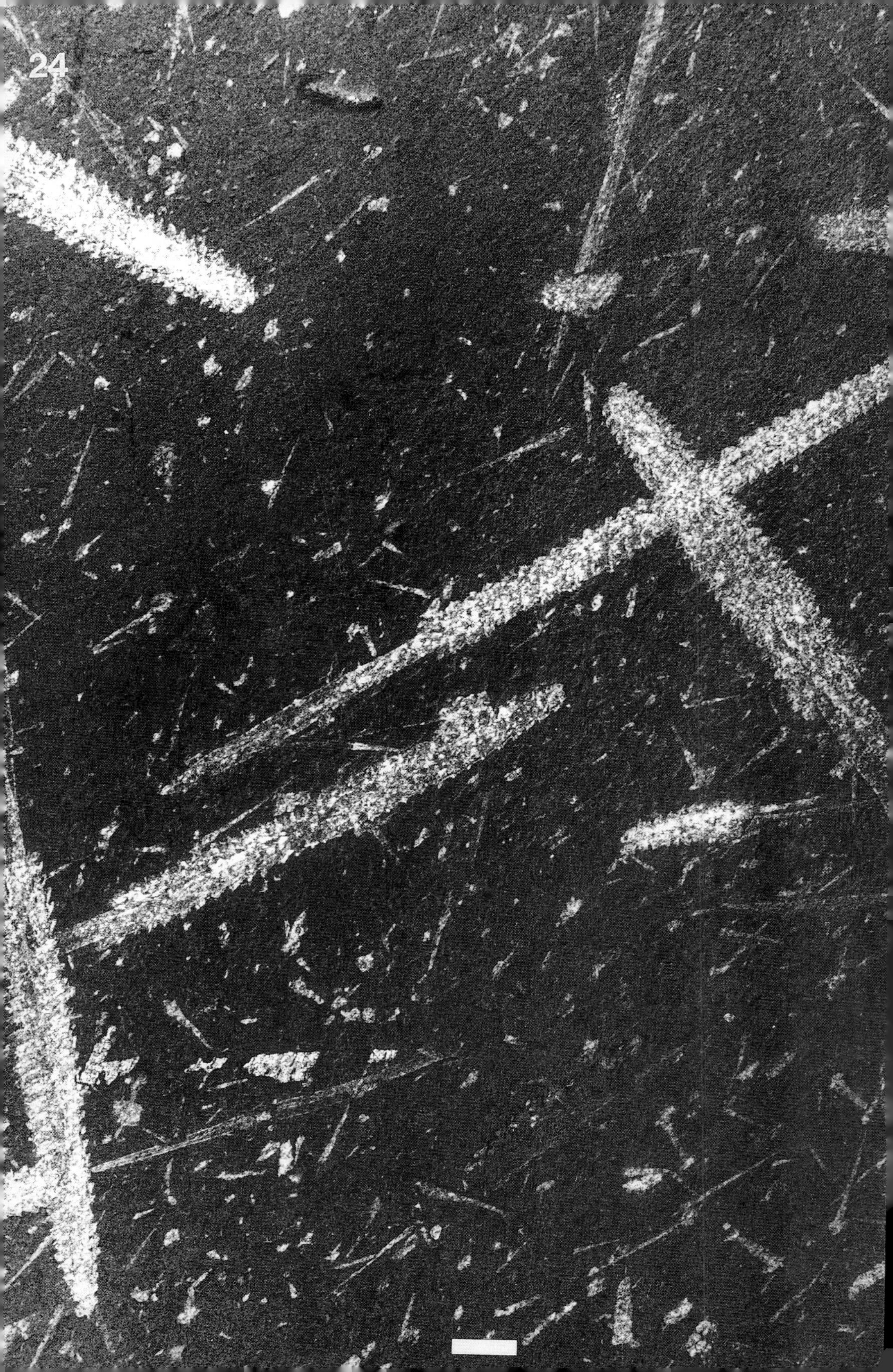
24

25

26

27

28

29

30
31

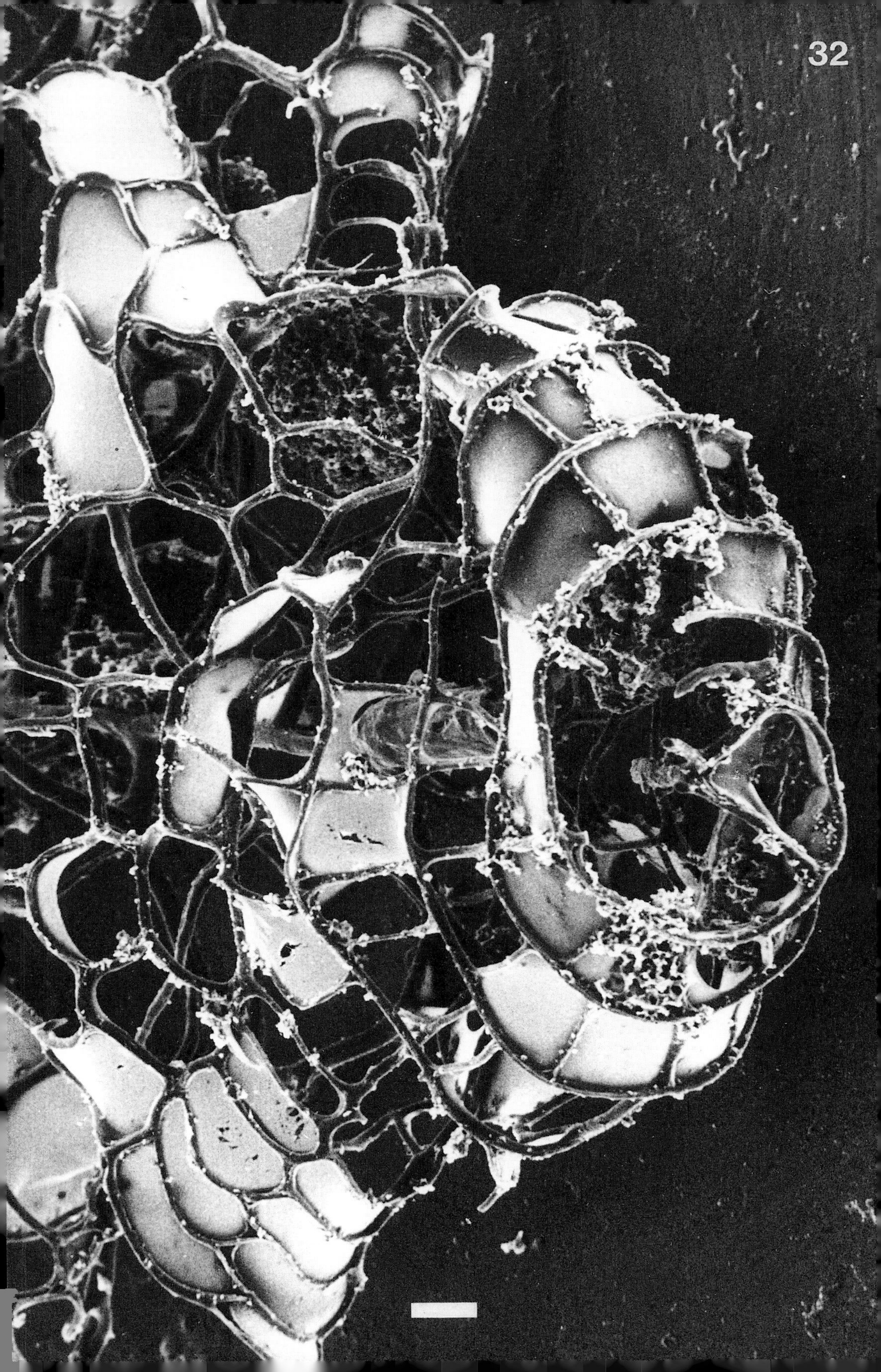
32

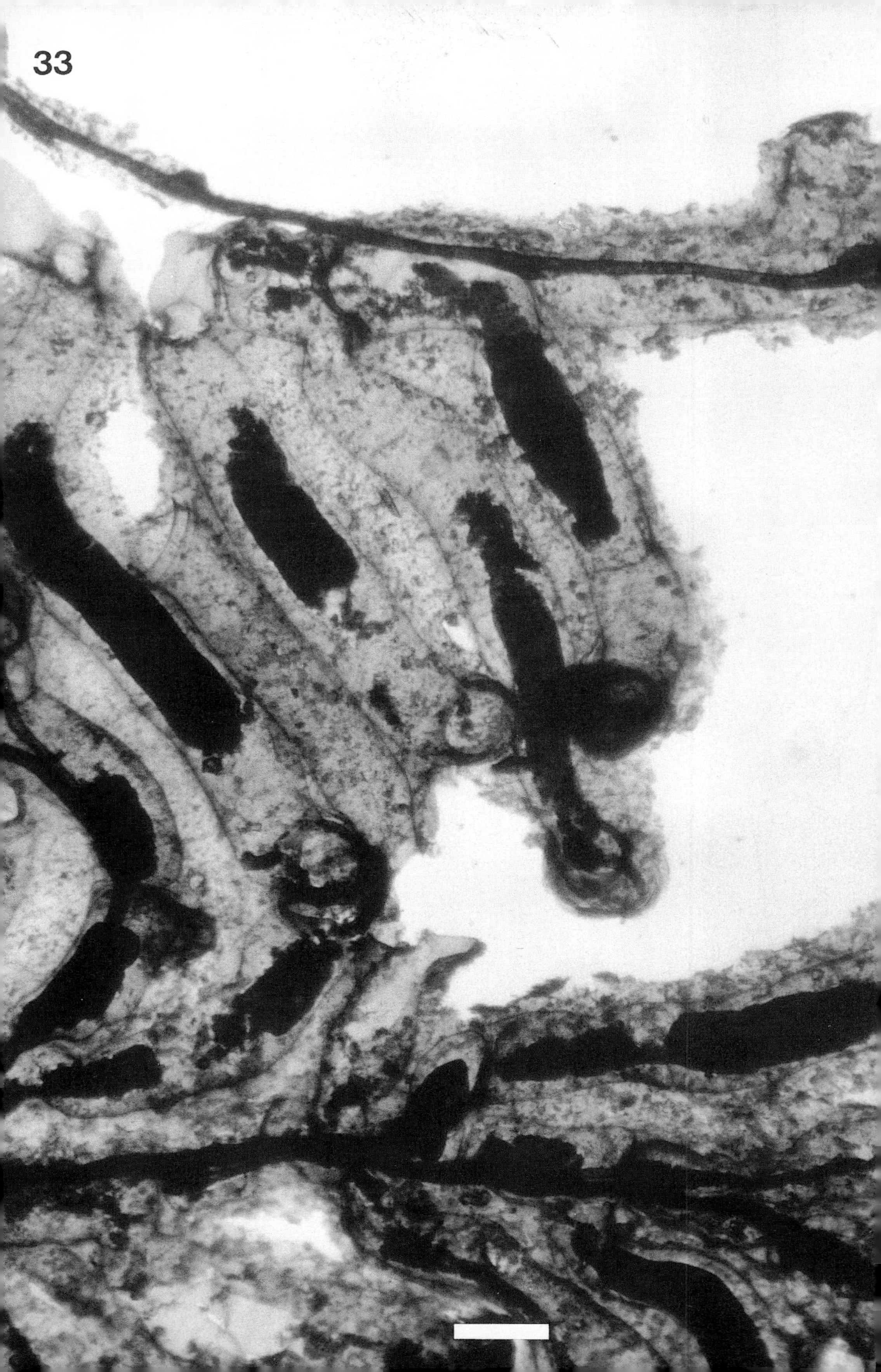
33

34

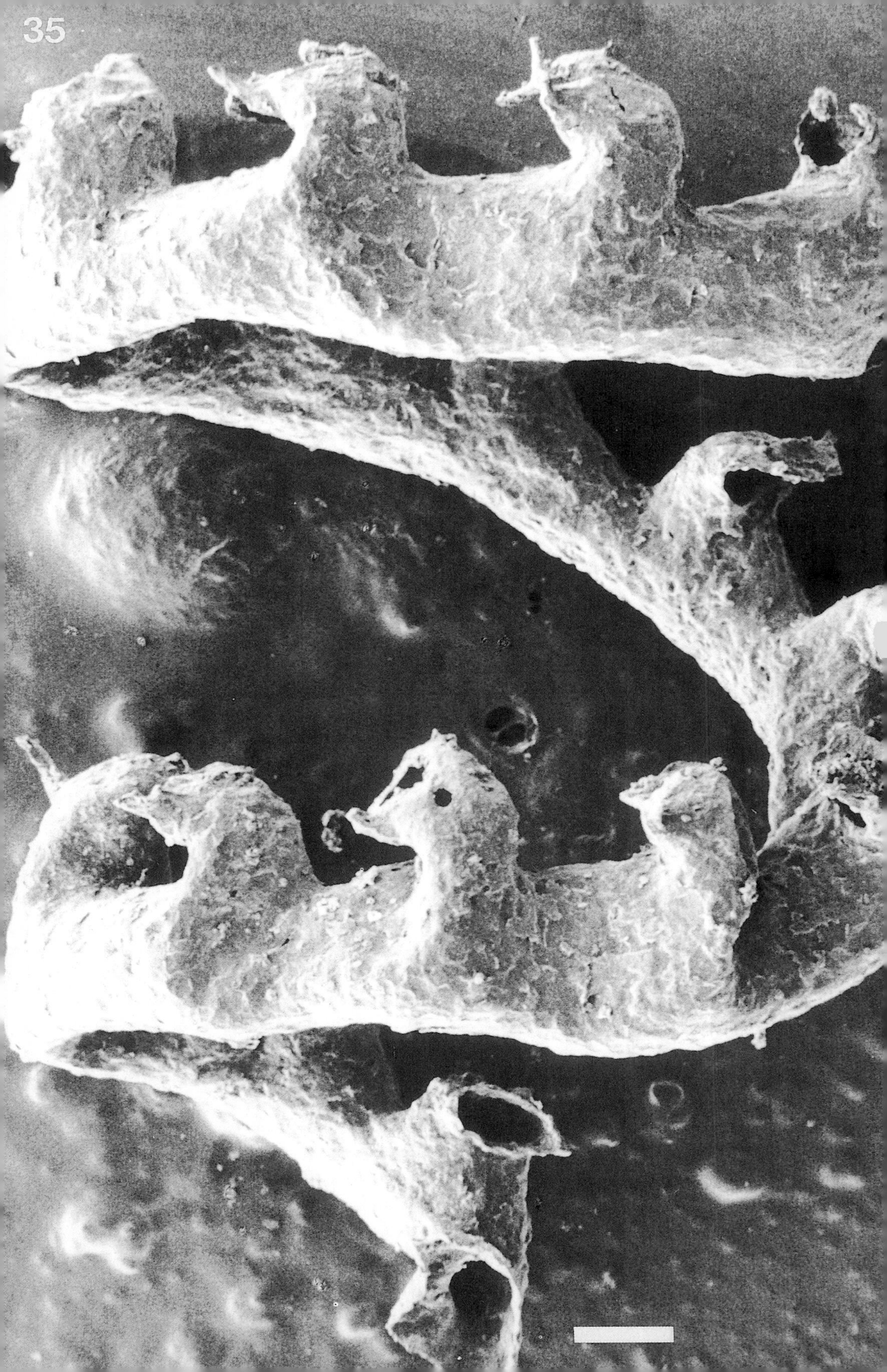
35

36

37
38

39

40

41

42

43

44

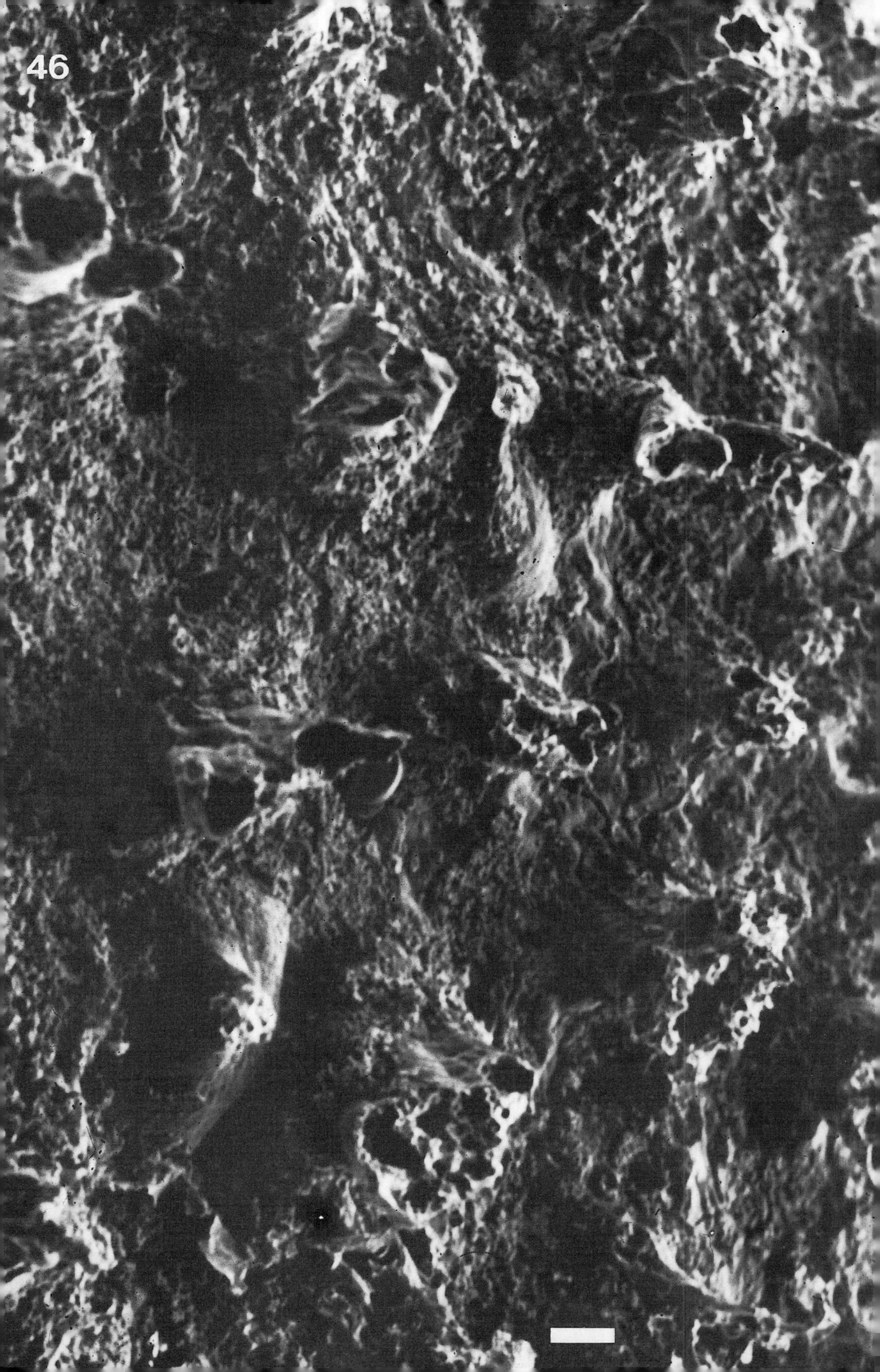
46

47

48

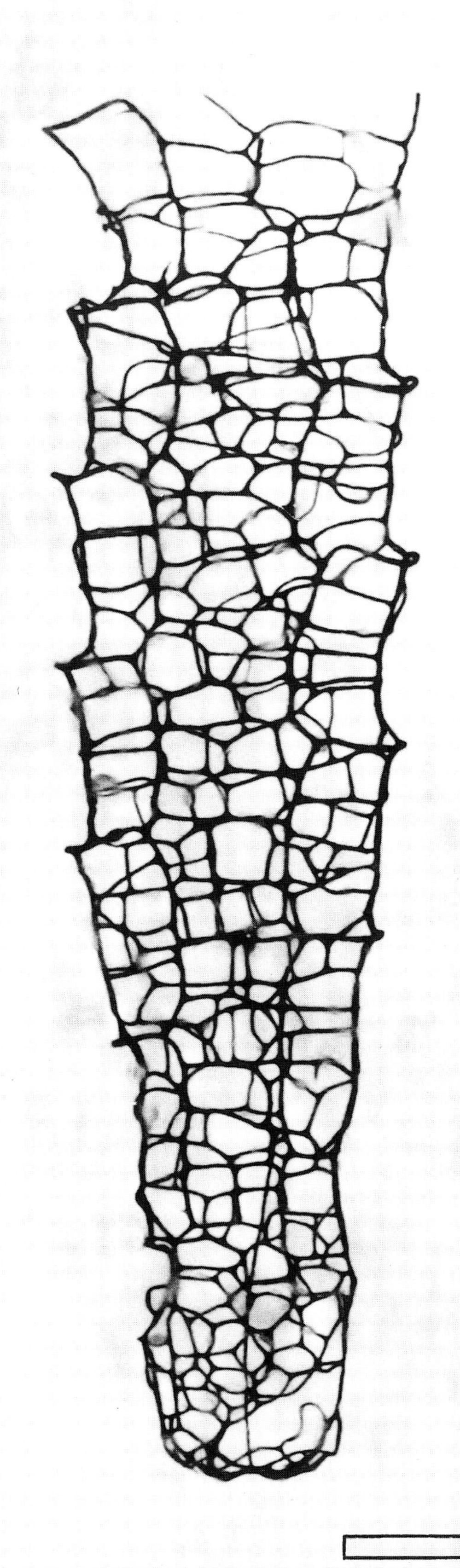

49

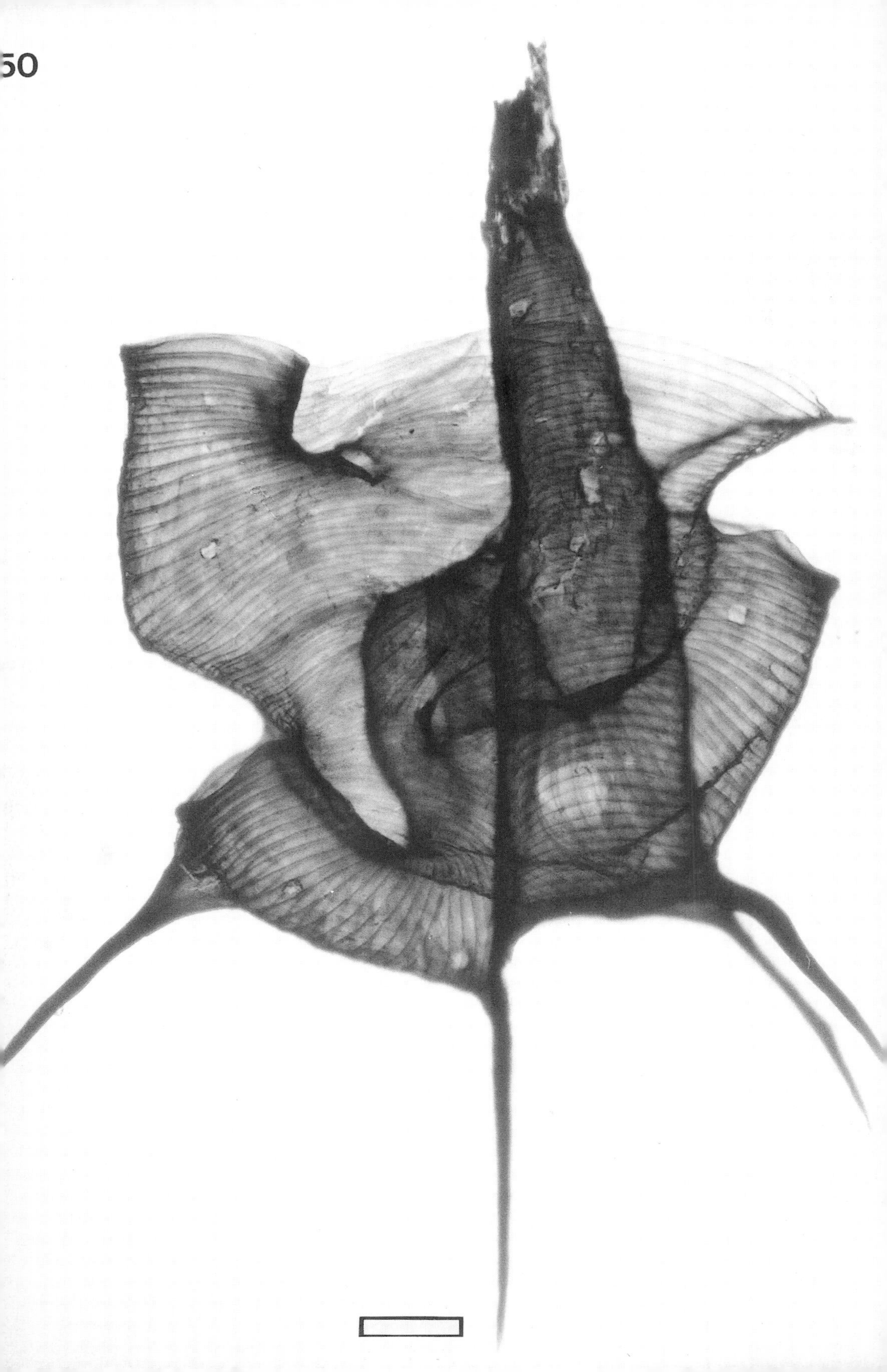
50

51

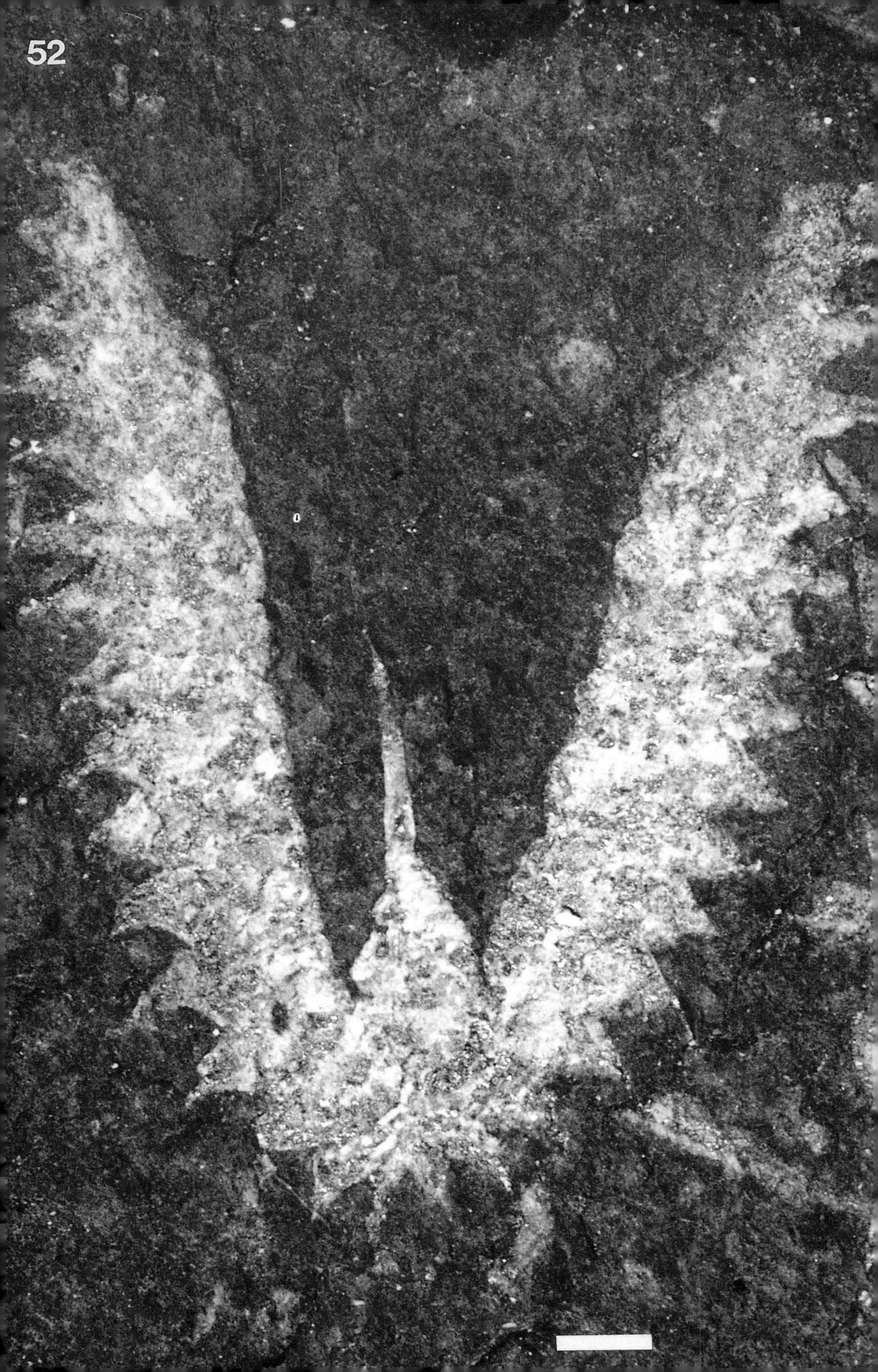
52

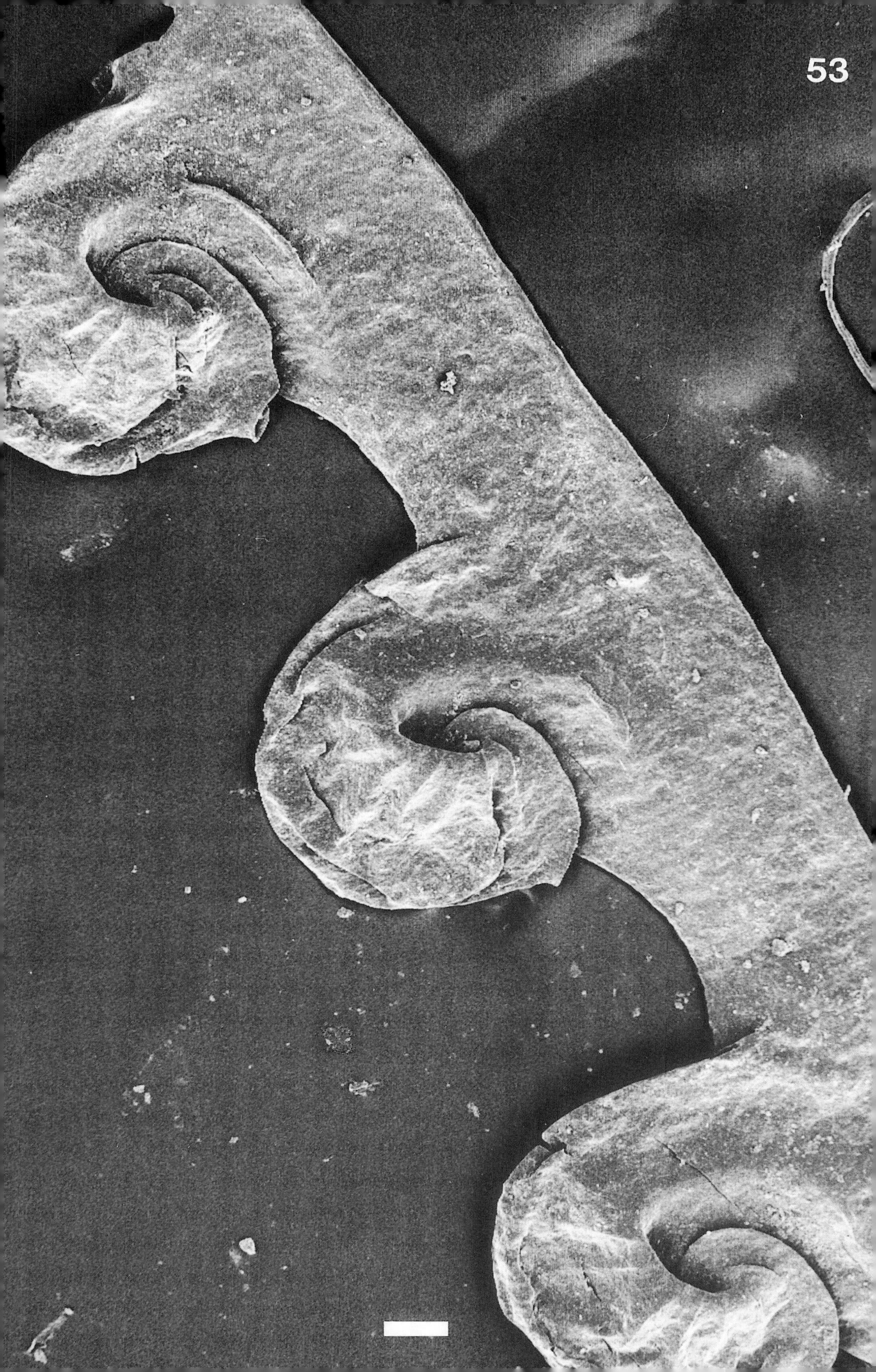
53

54

55

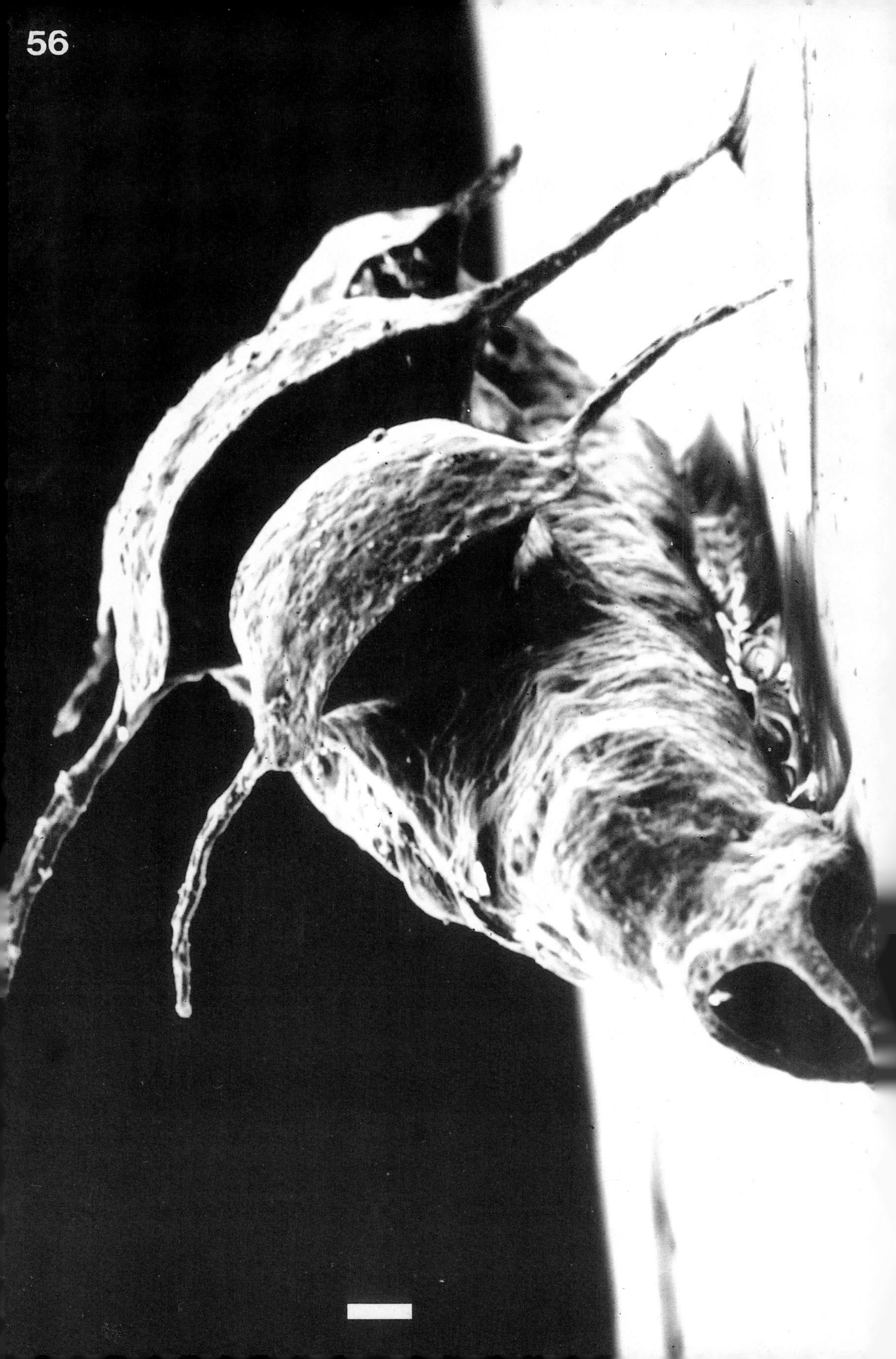
56

57

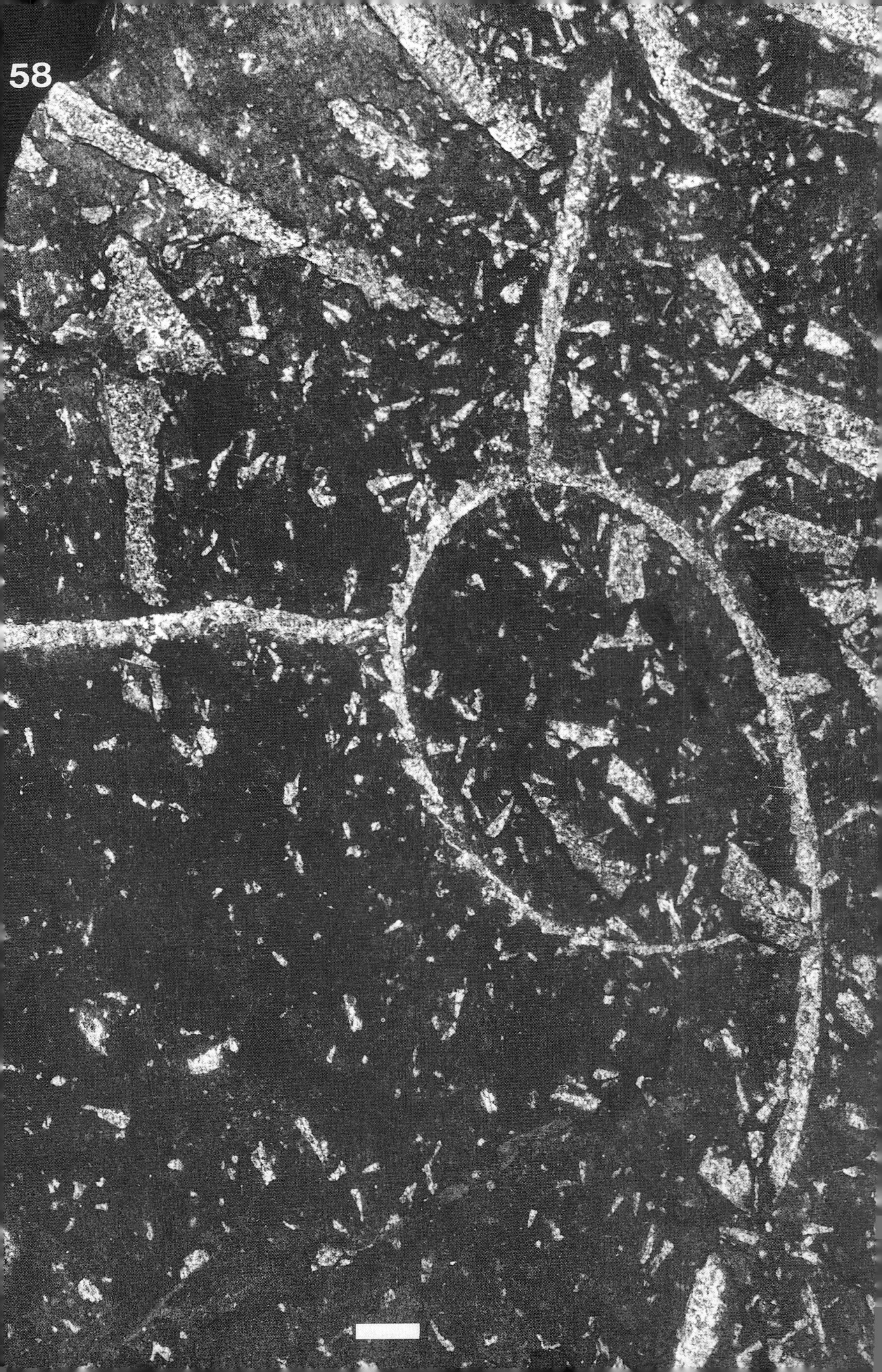
58

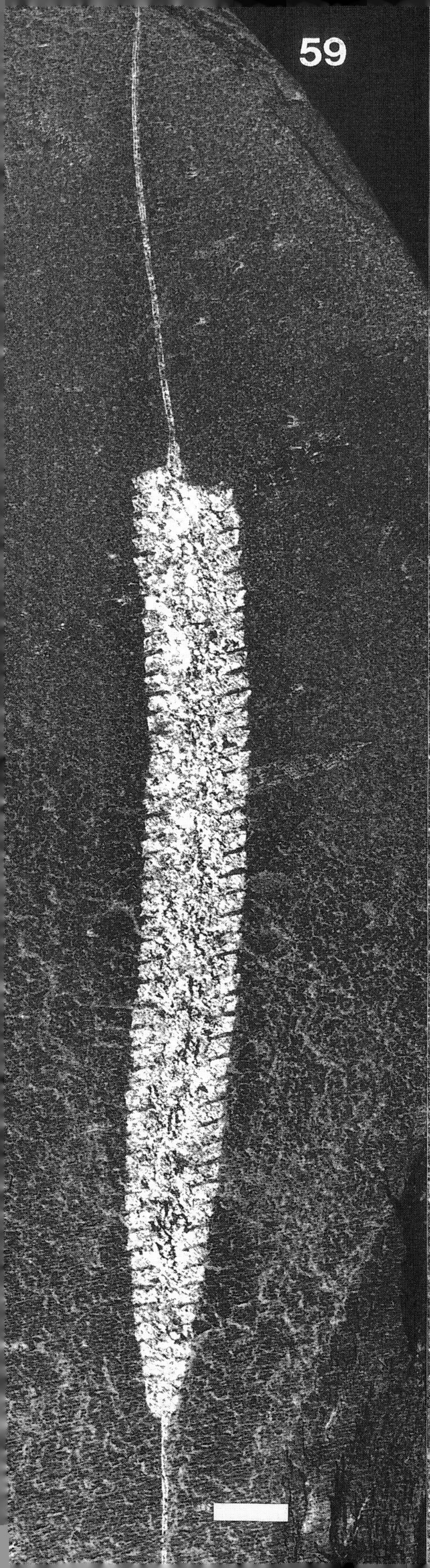

59

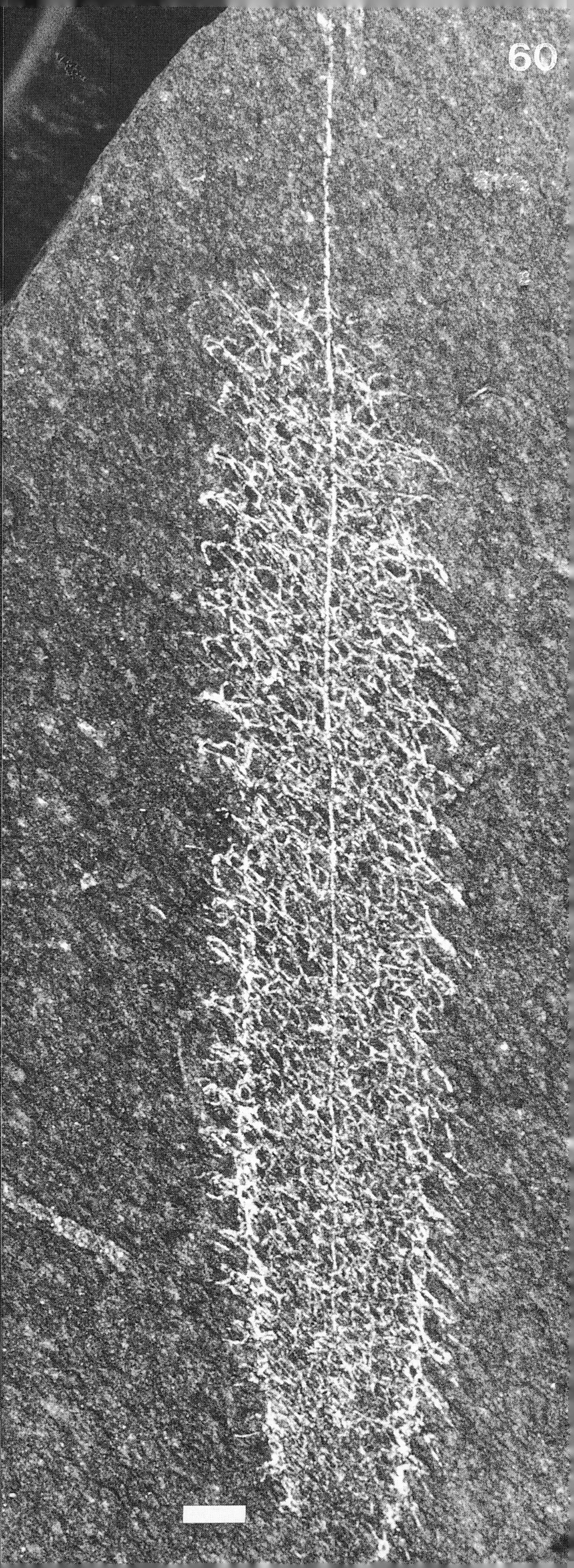

60

61

62

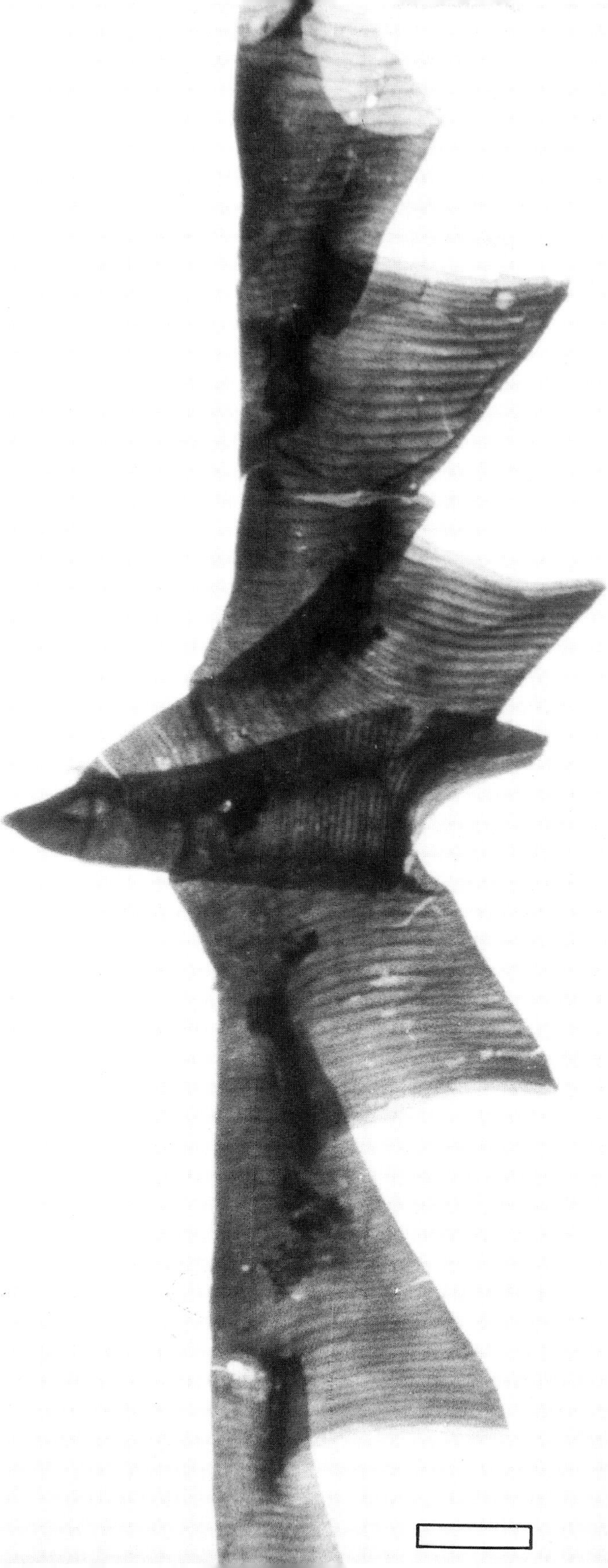

63

64

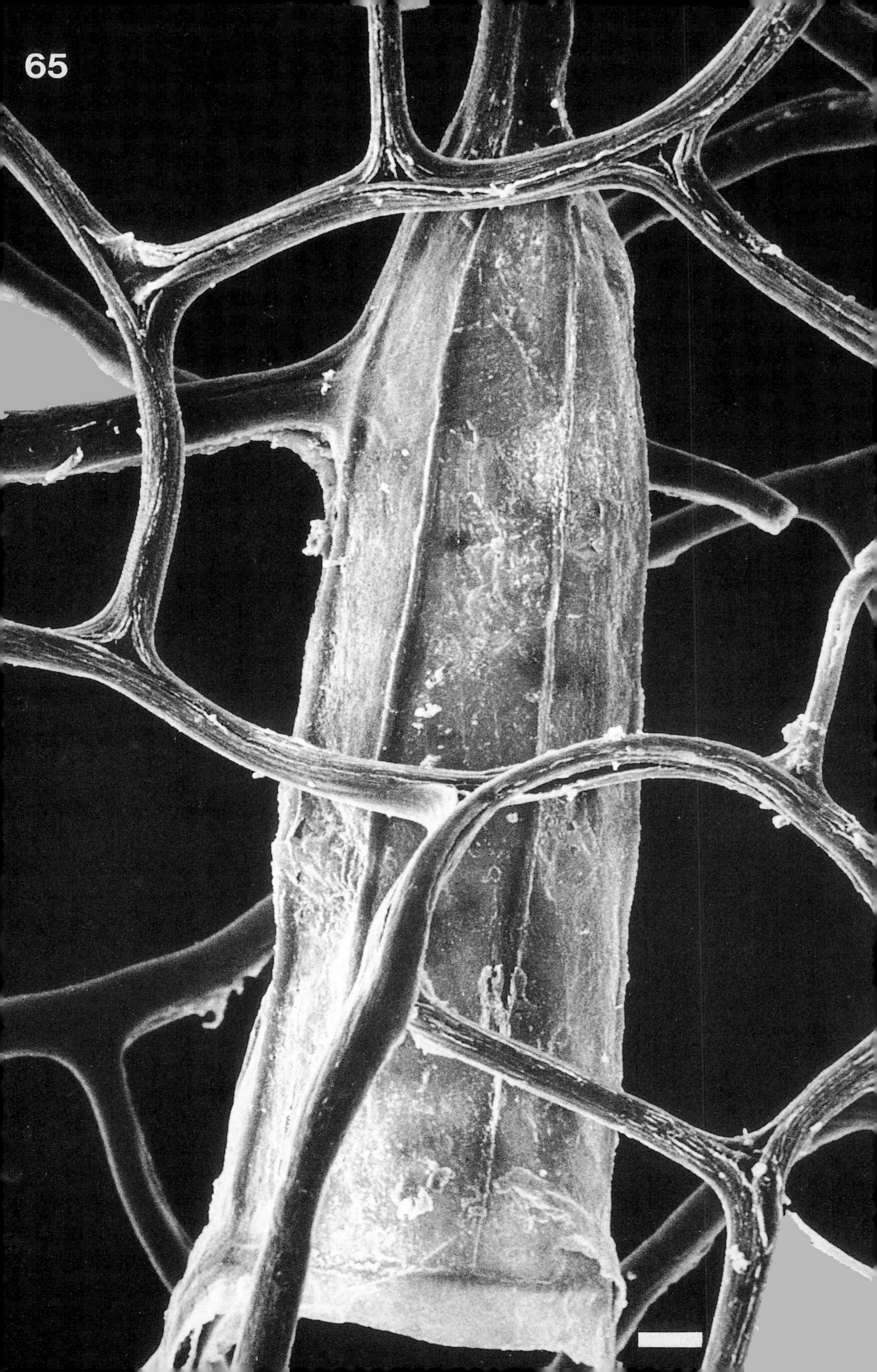
65

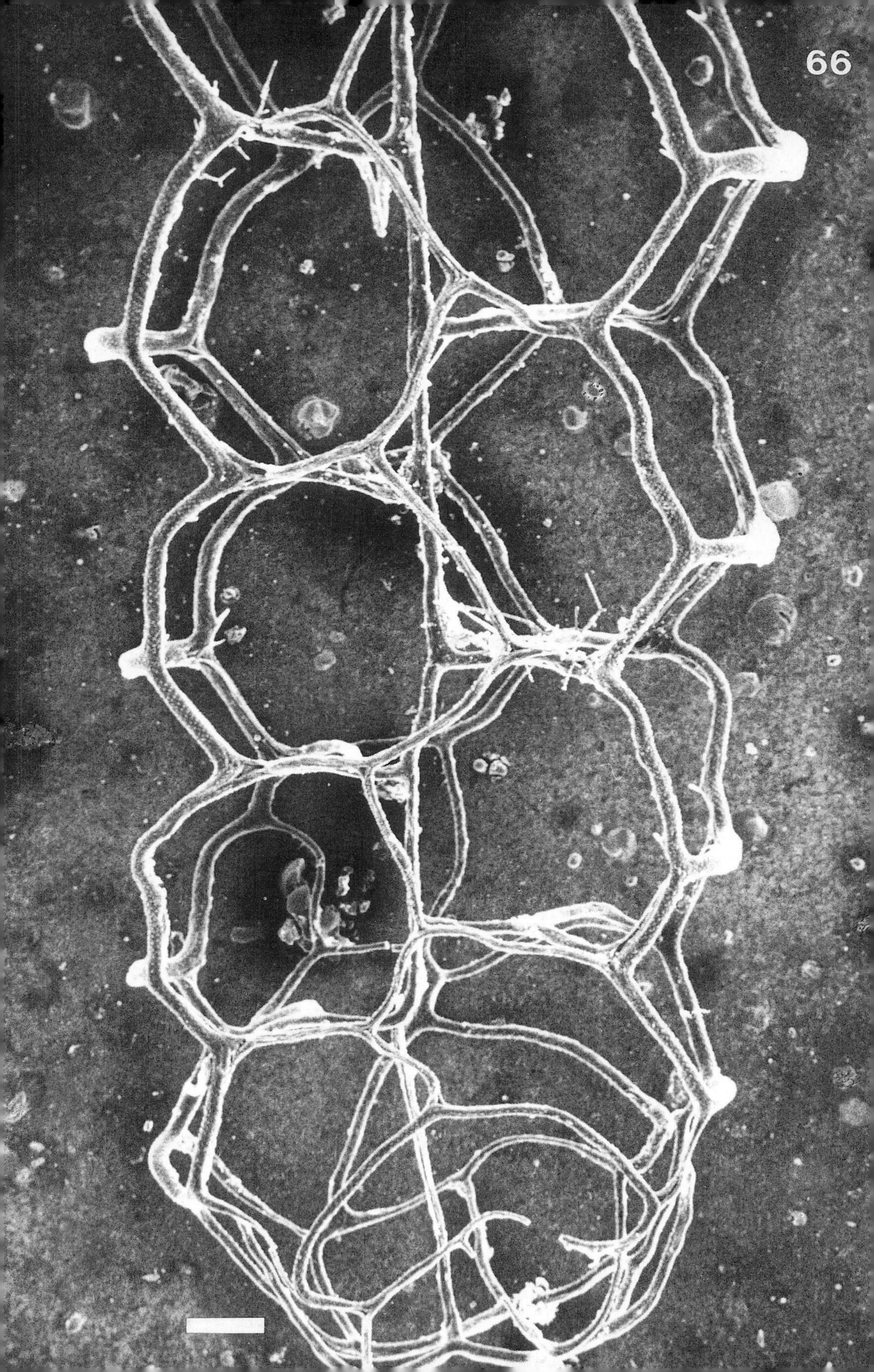
66

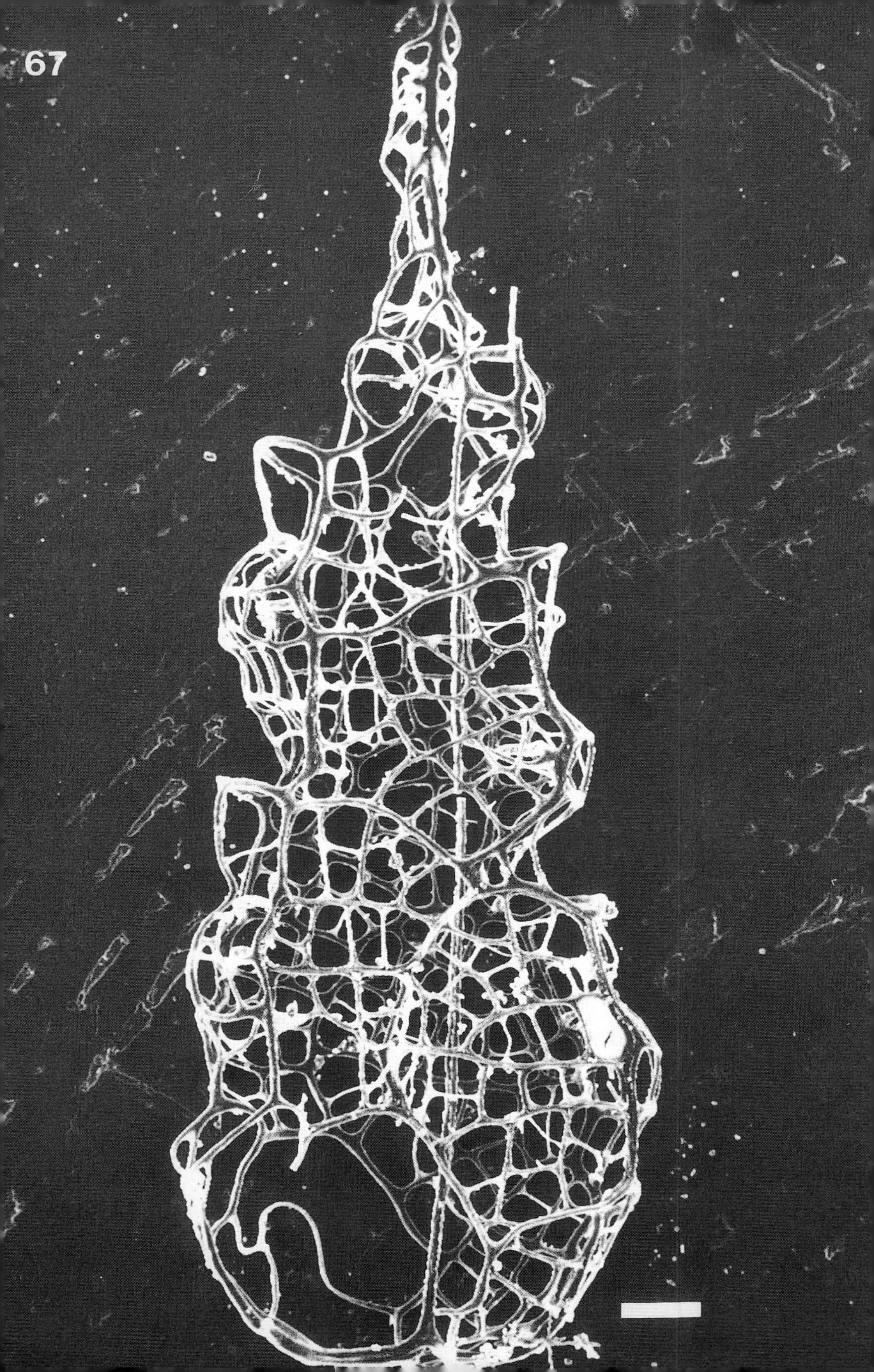
67

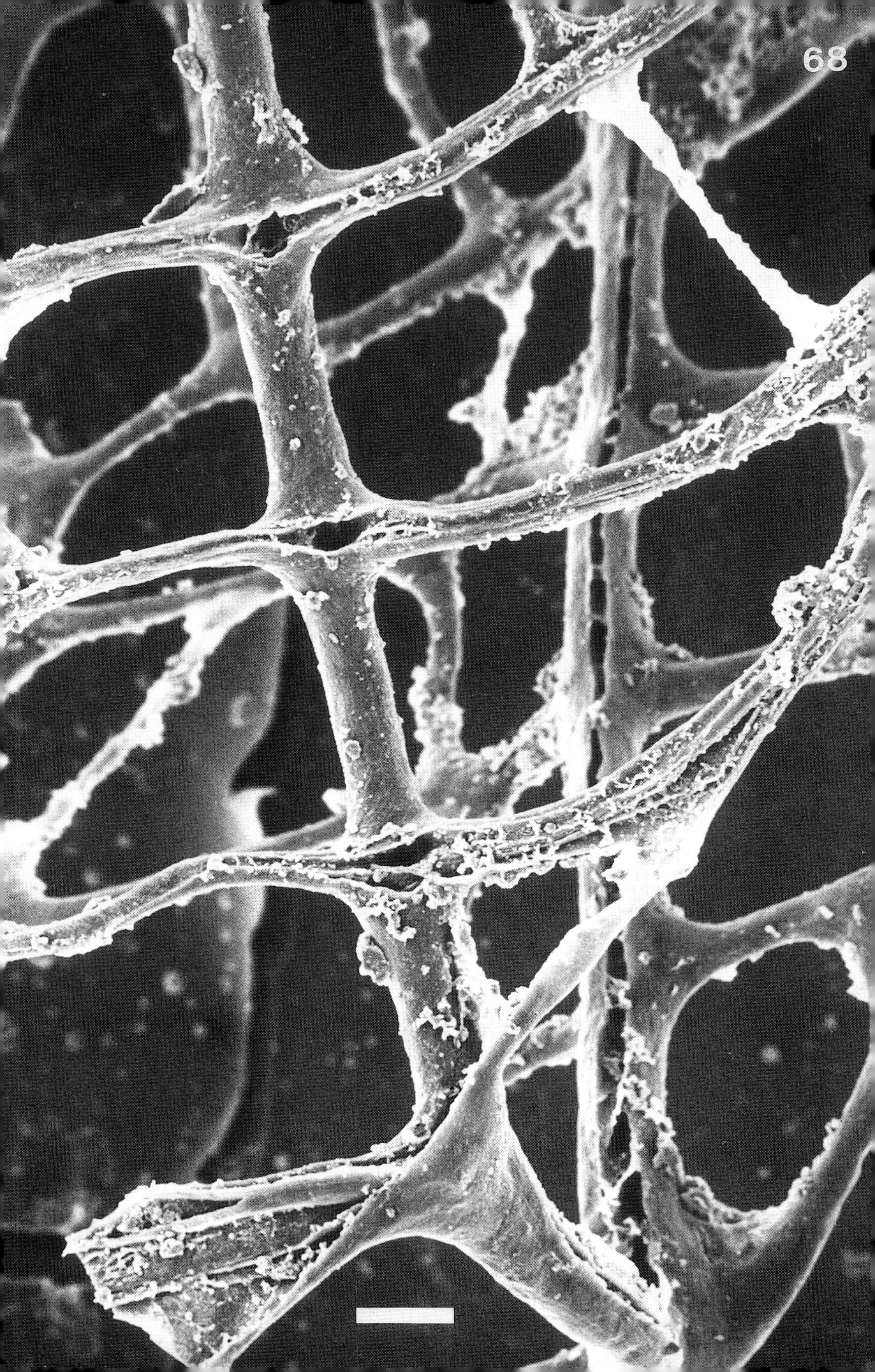
68

69

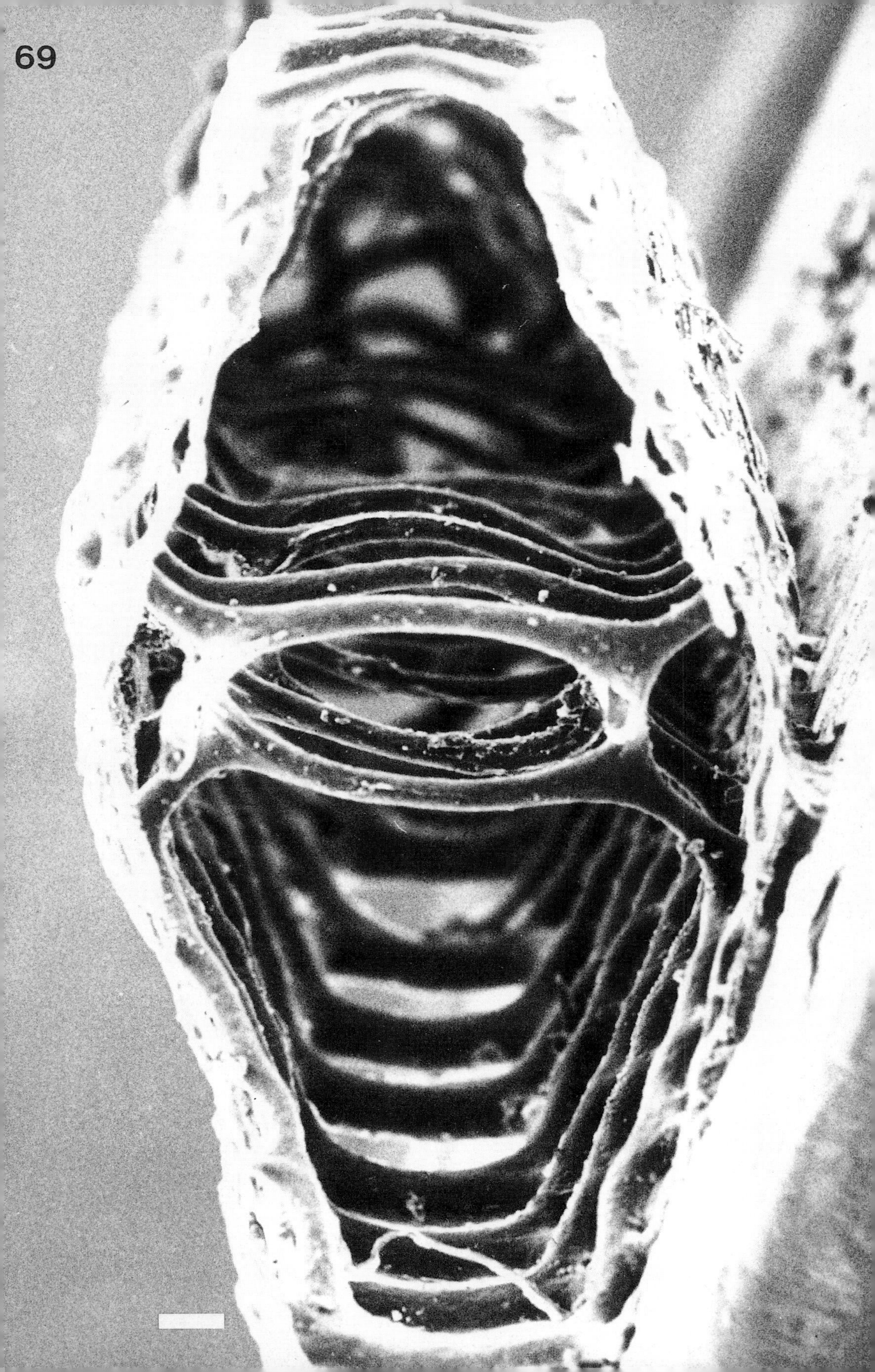

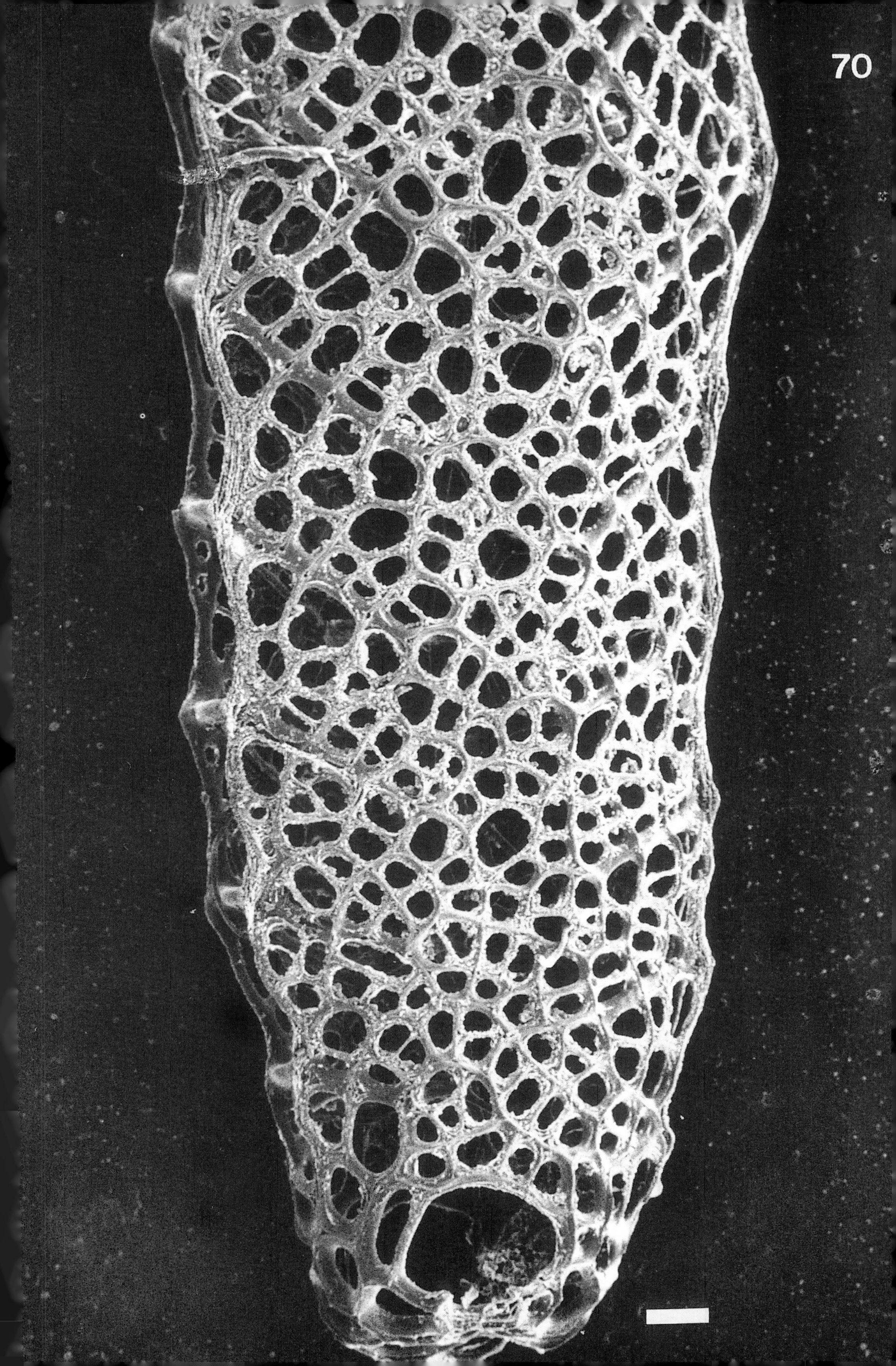

70

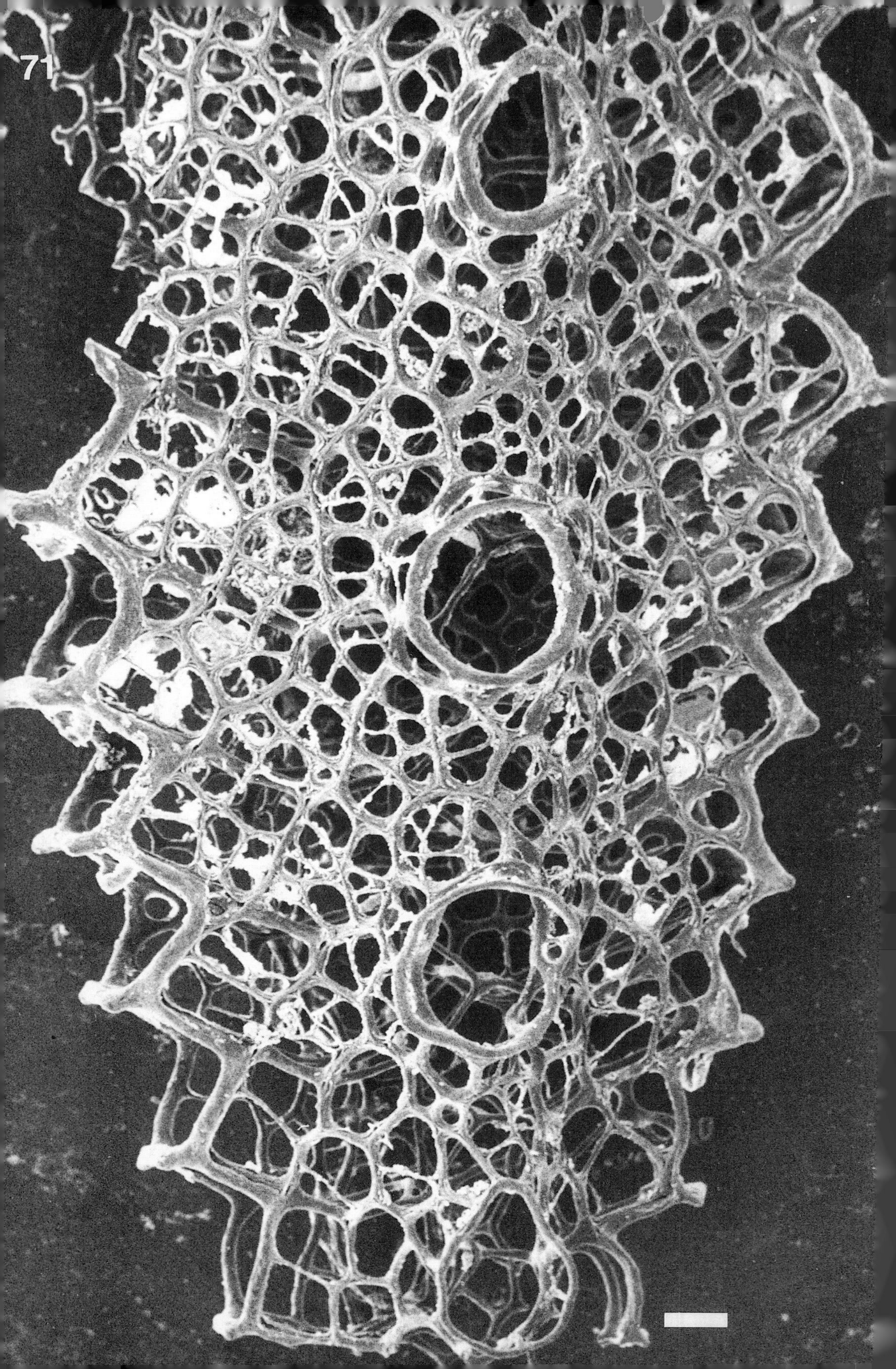
71

72
73

74

75

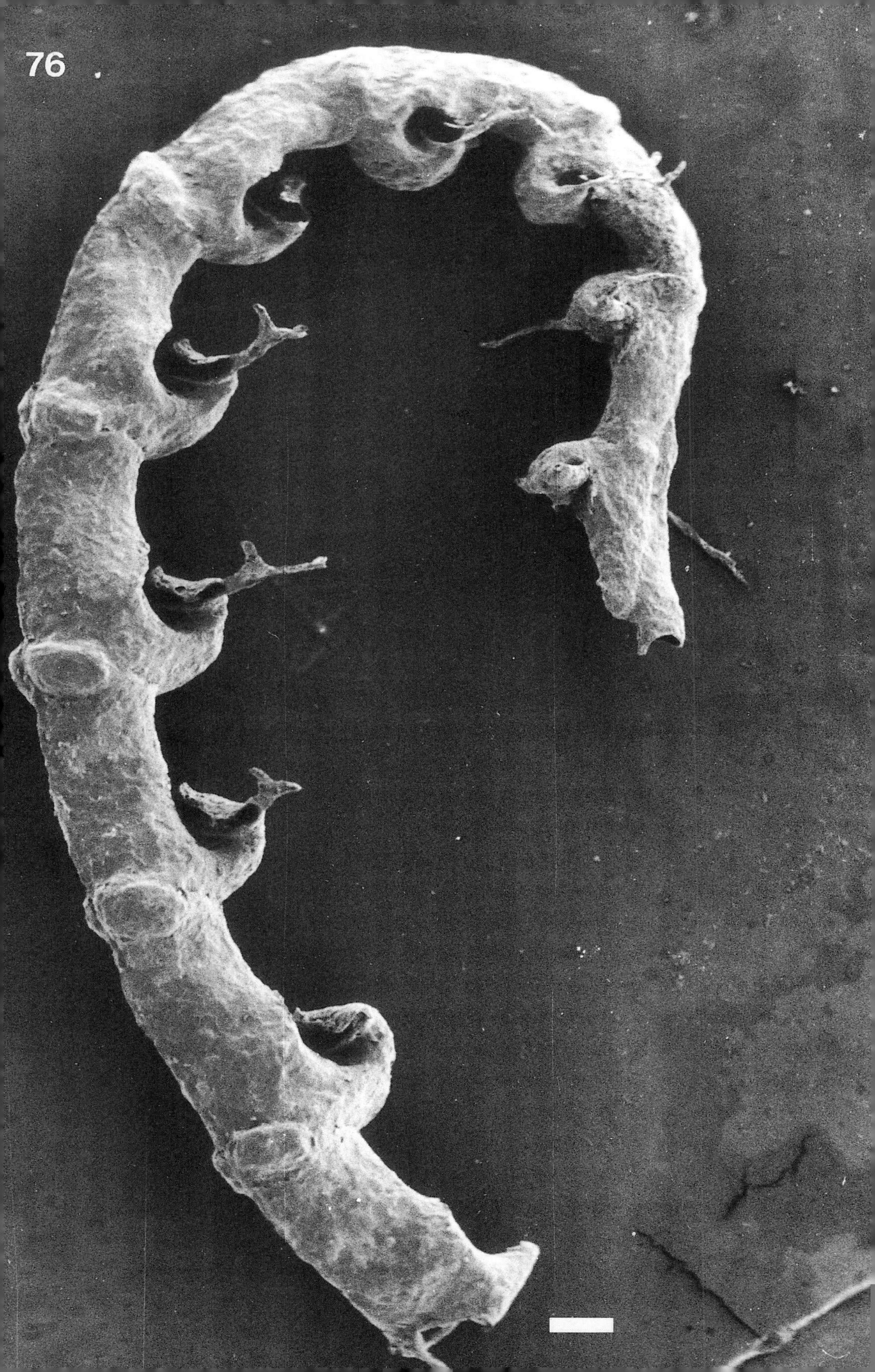
76

77

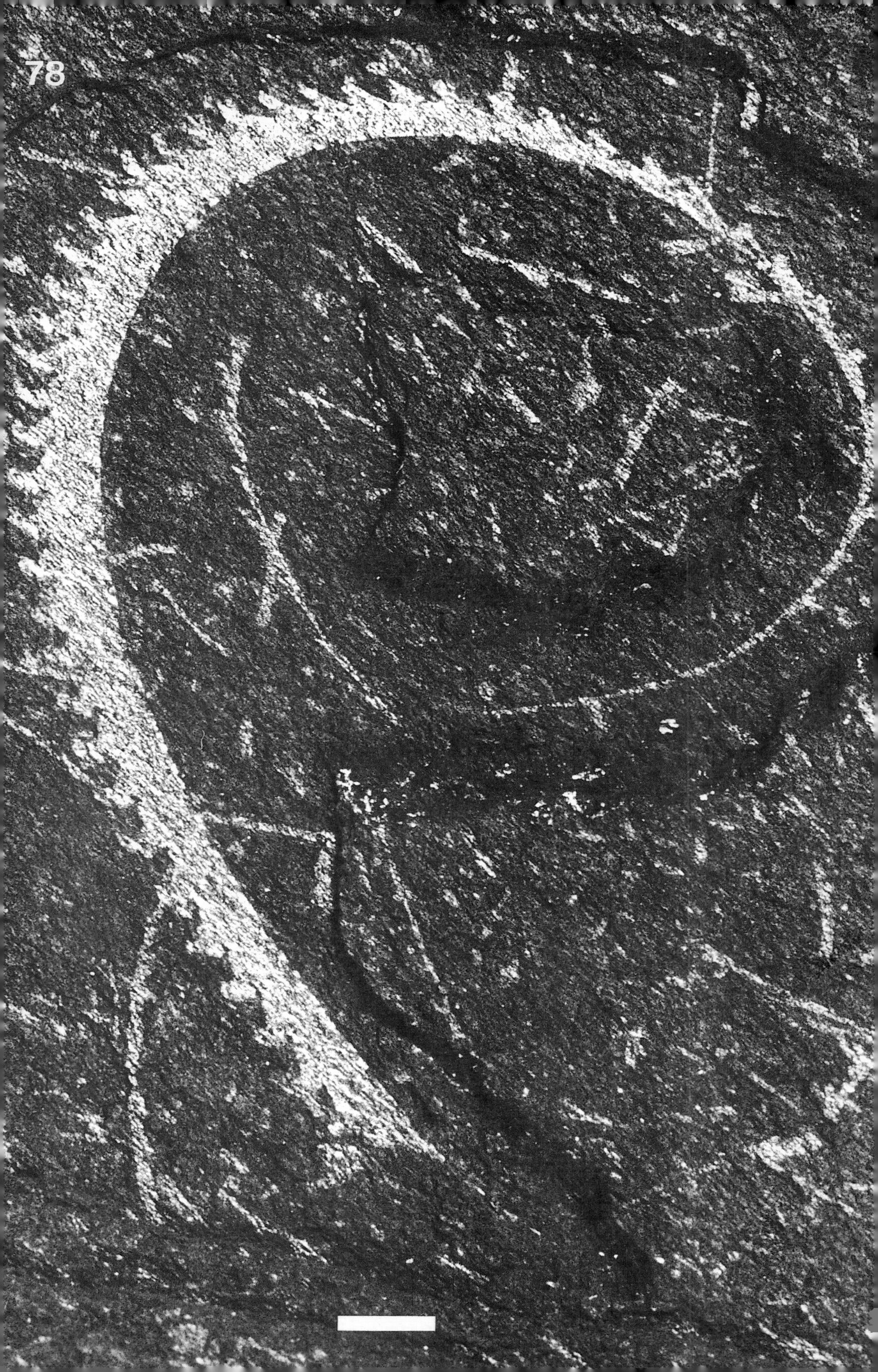
78

79

80

81

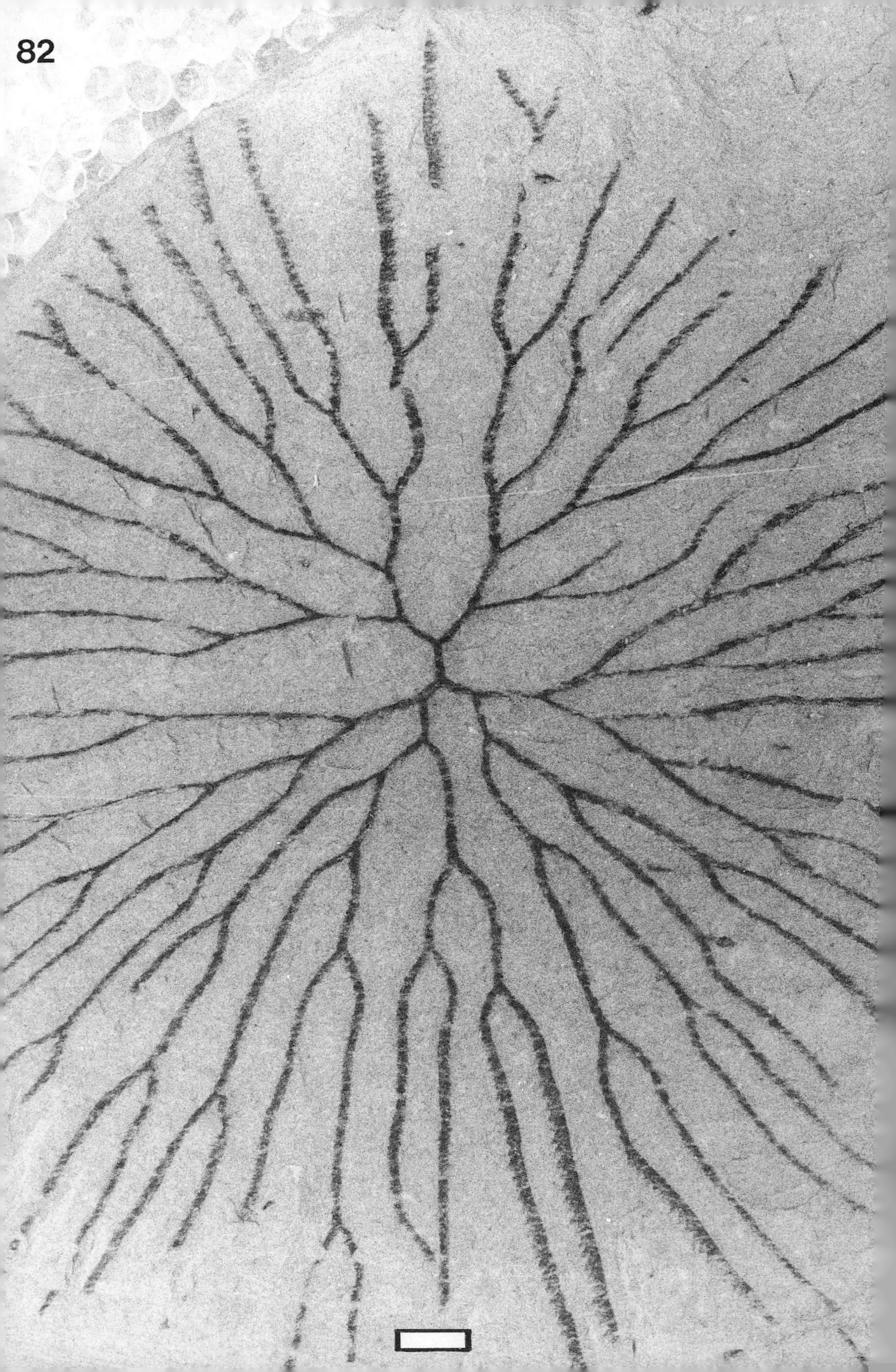
82

83

84

85

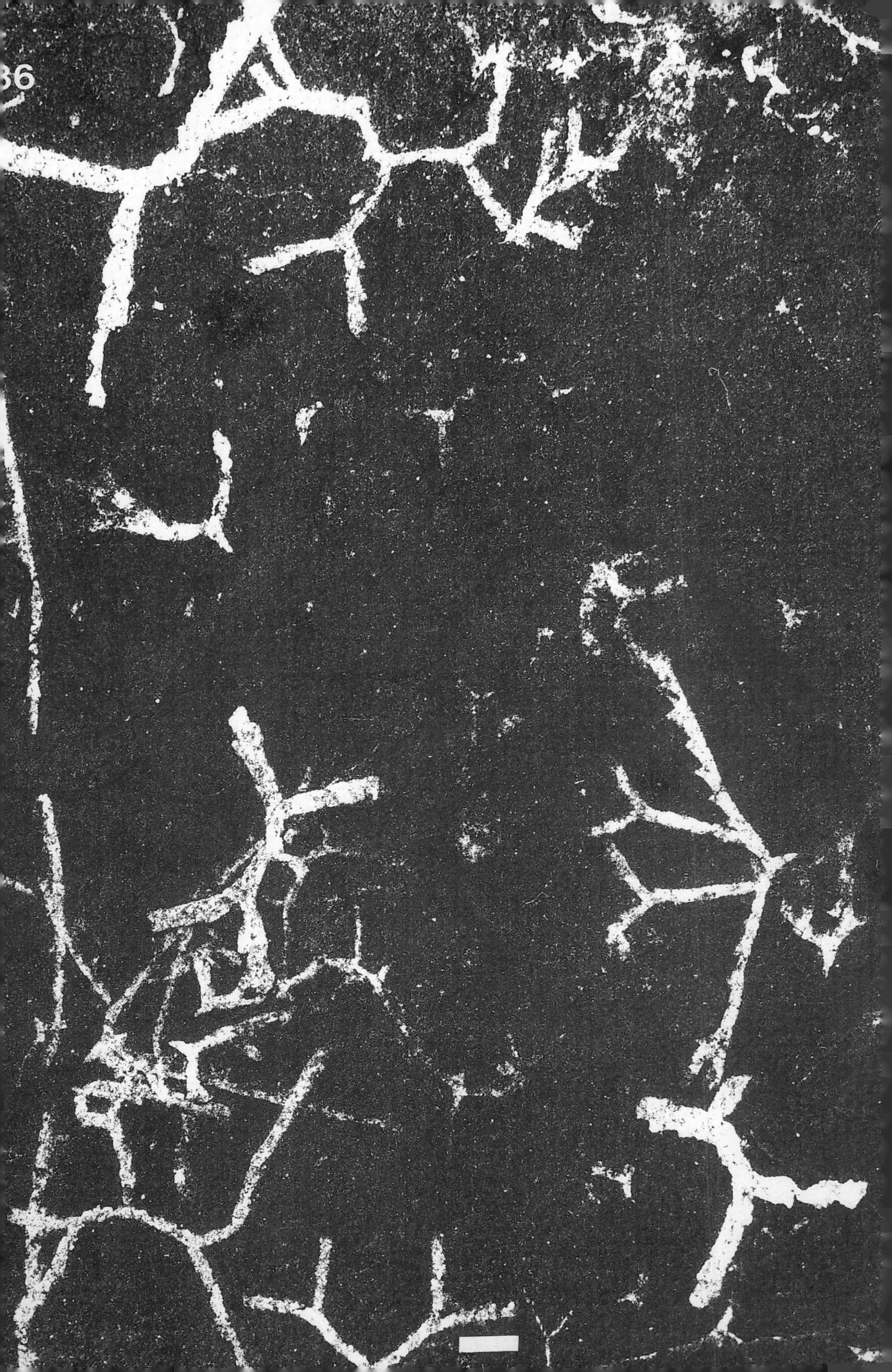
36

87

88

89

90

9

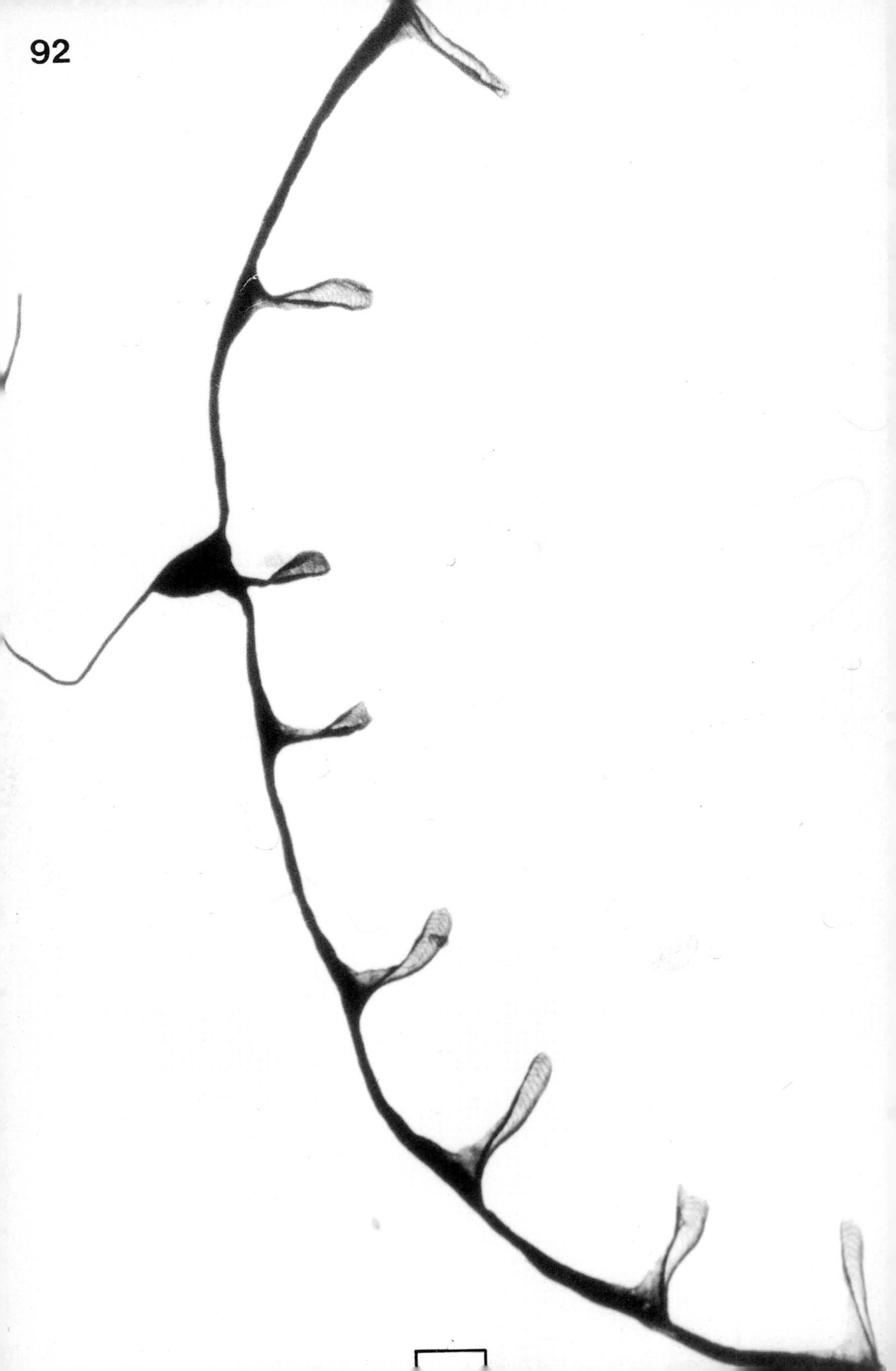
92

93

94

95
96

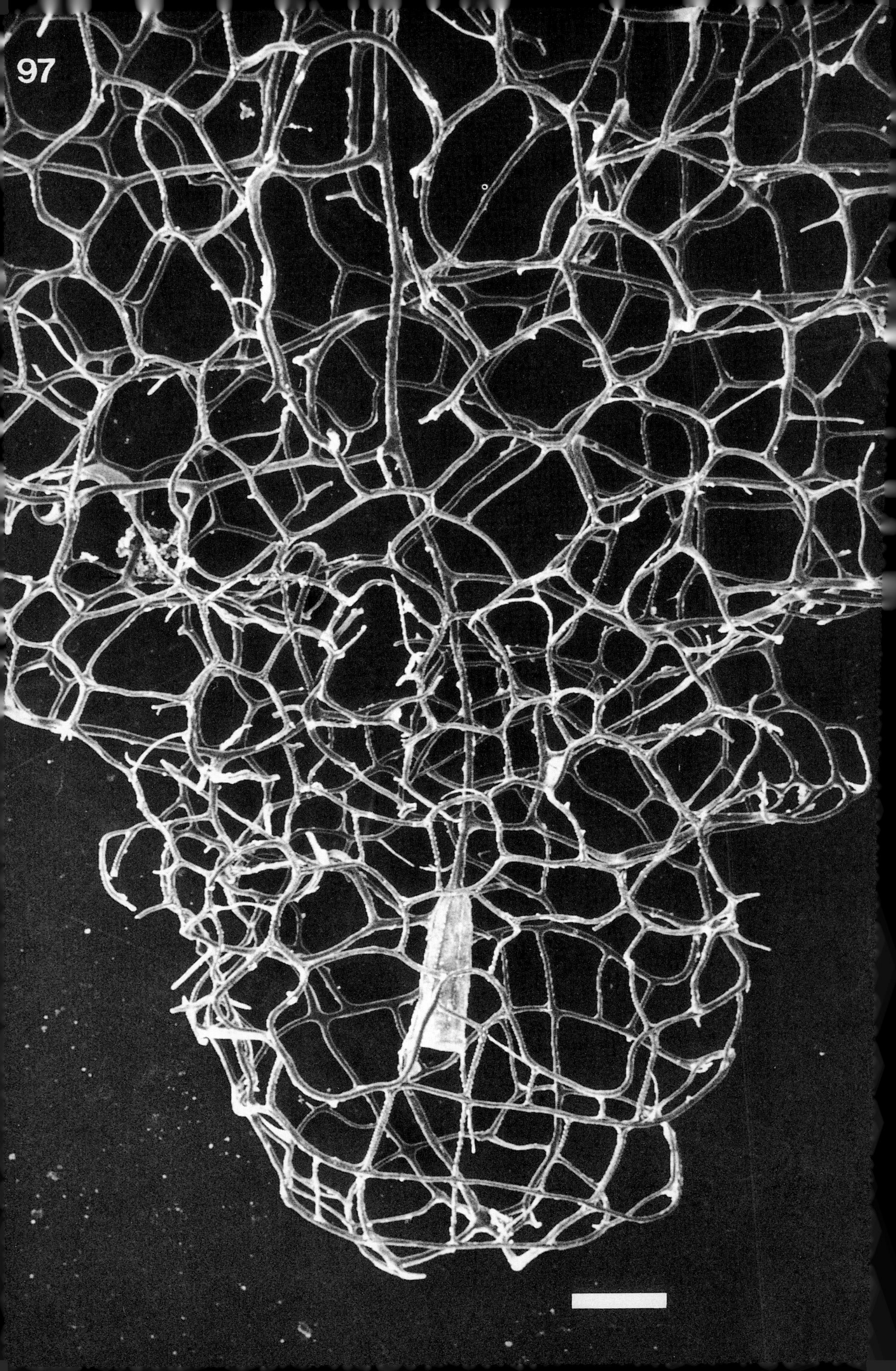
97

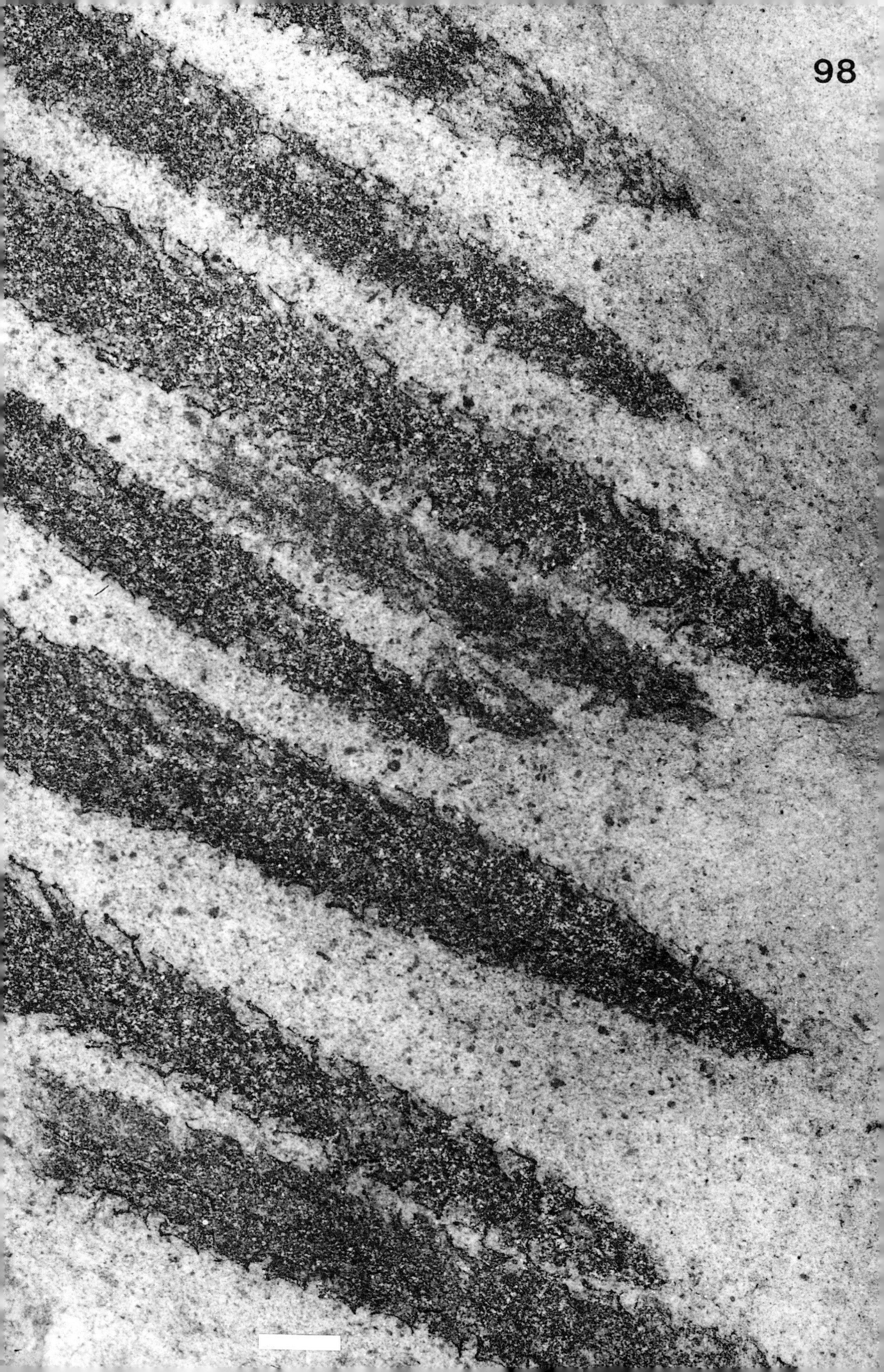
98

99
100

101

102

103
104

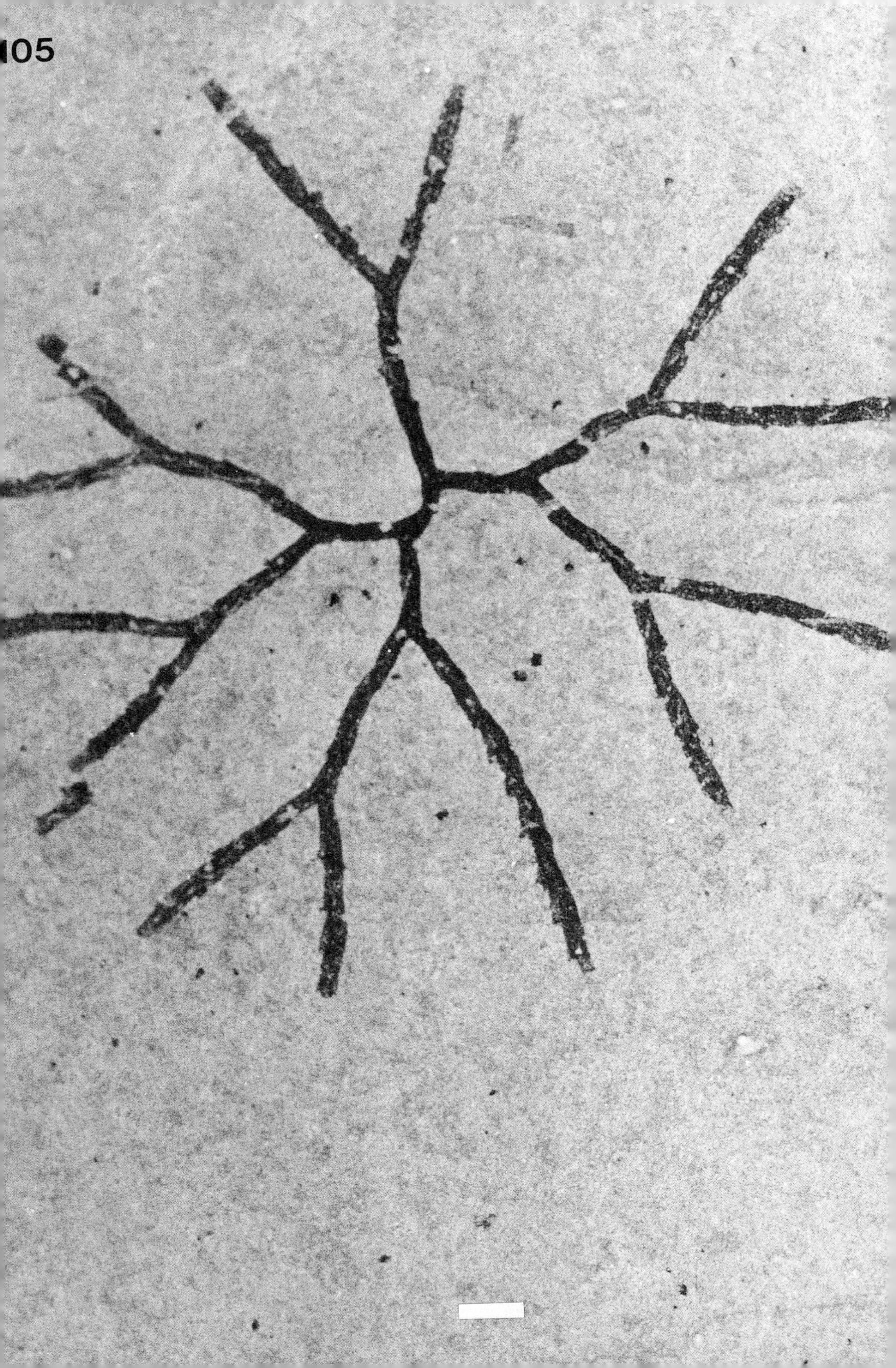
105

106
107

108

109

110

A 32748

112

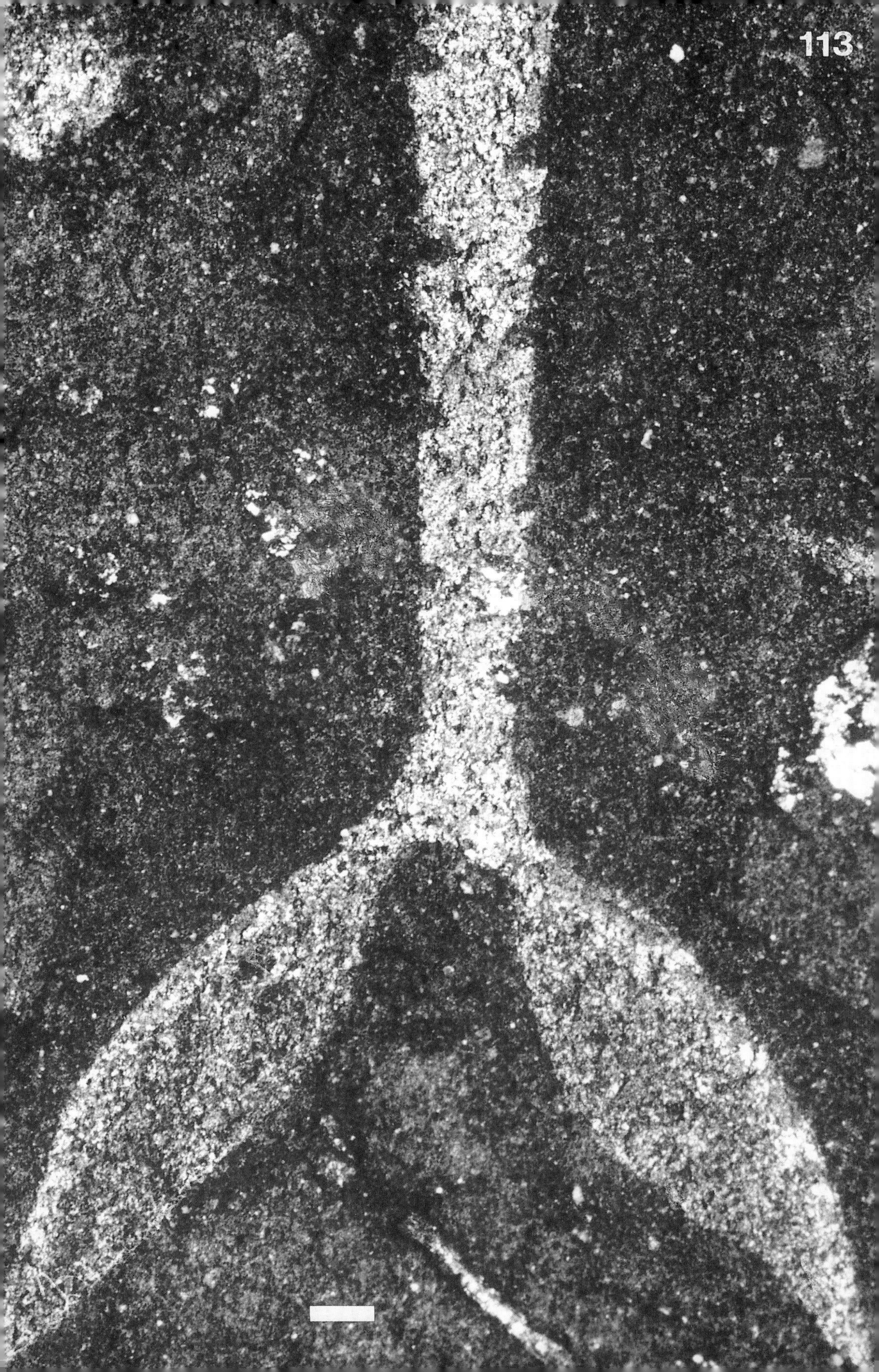
113

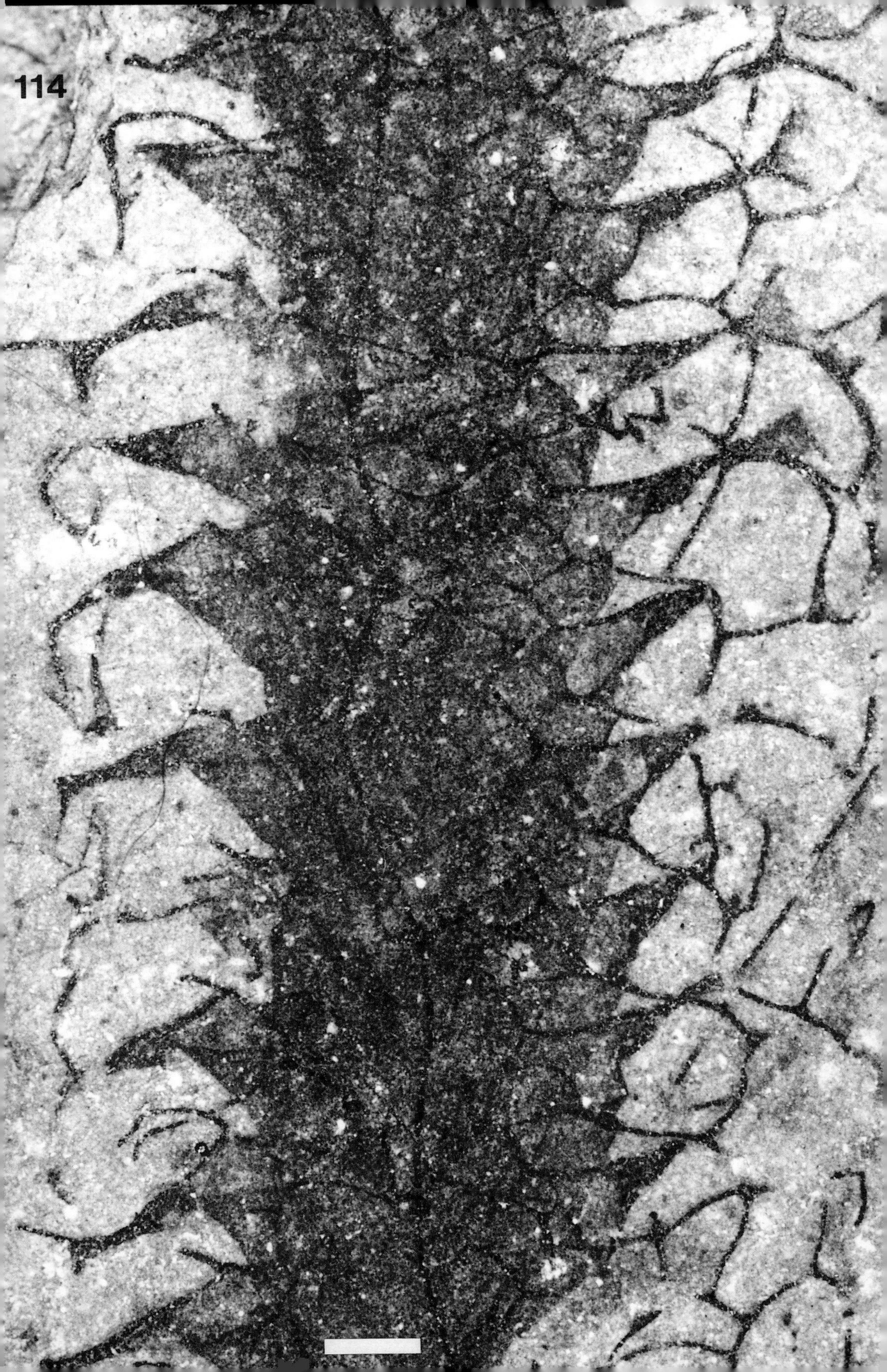
114

115

116

117

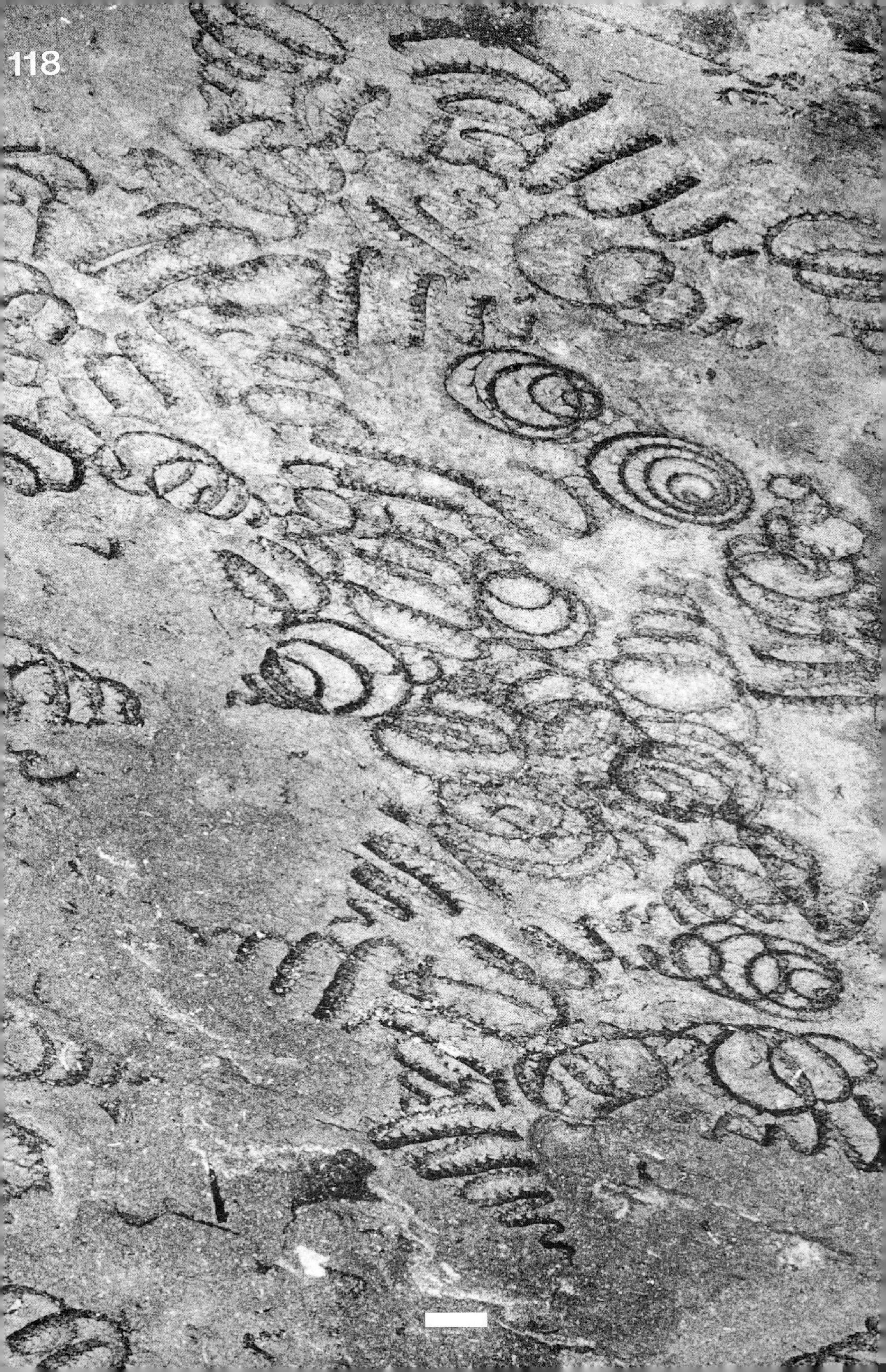
118

119

120

121

122

123

124

125

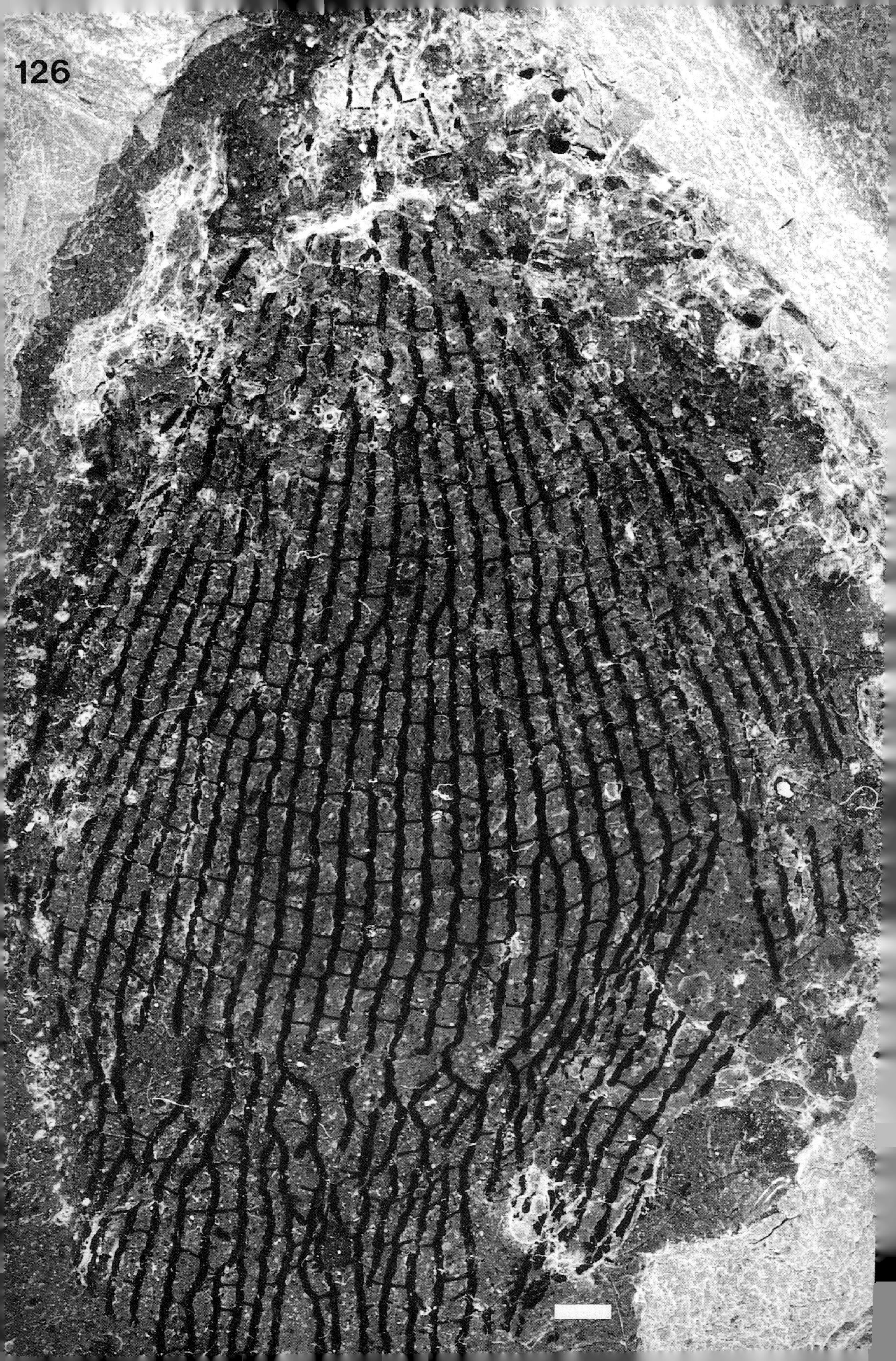
126

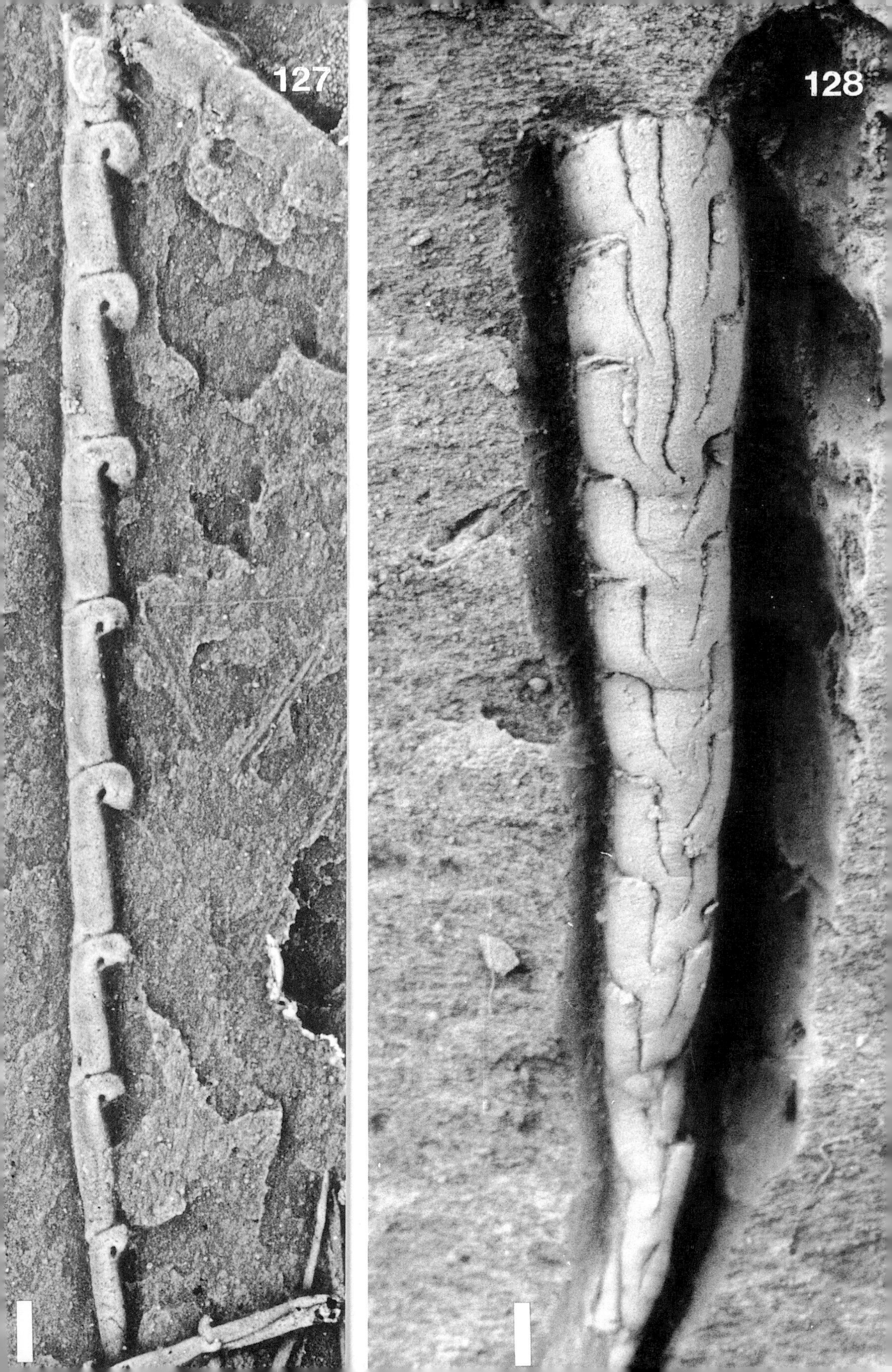
127
128

129
130

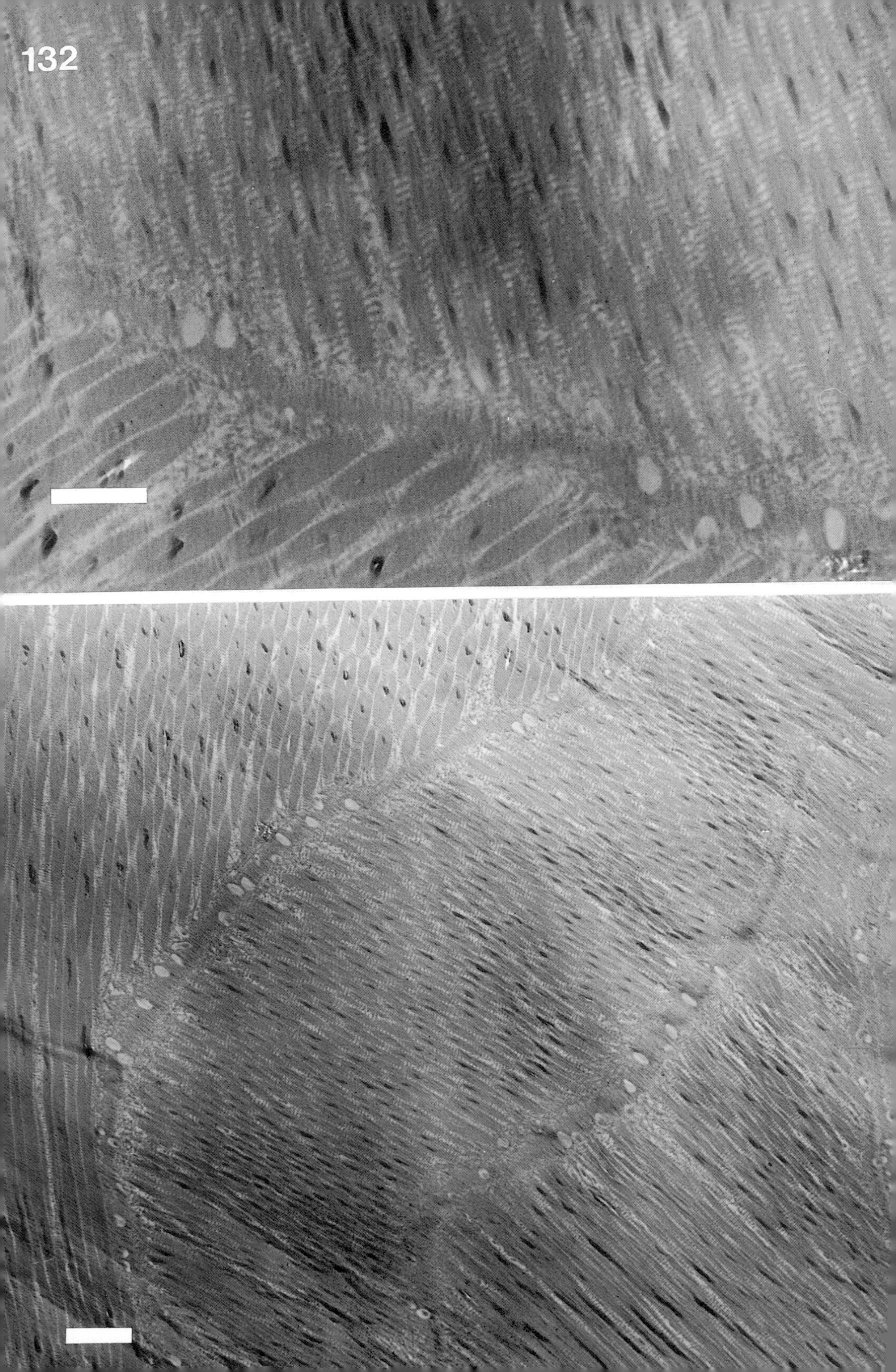
132

133

134

135

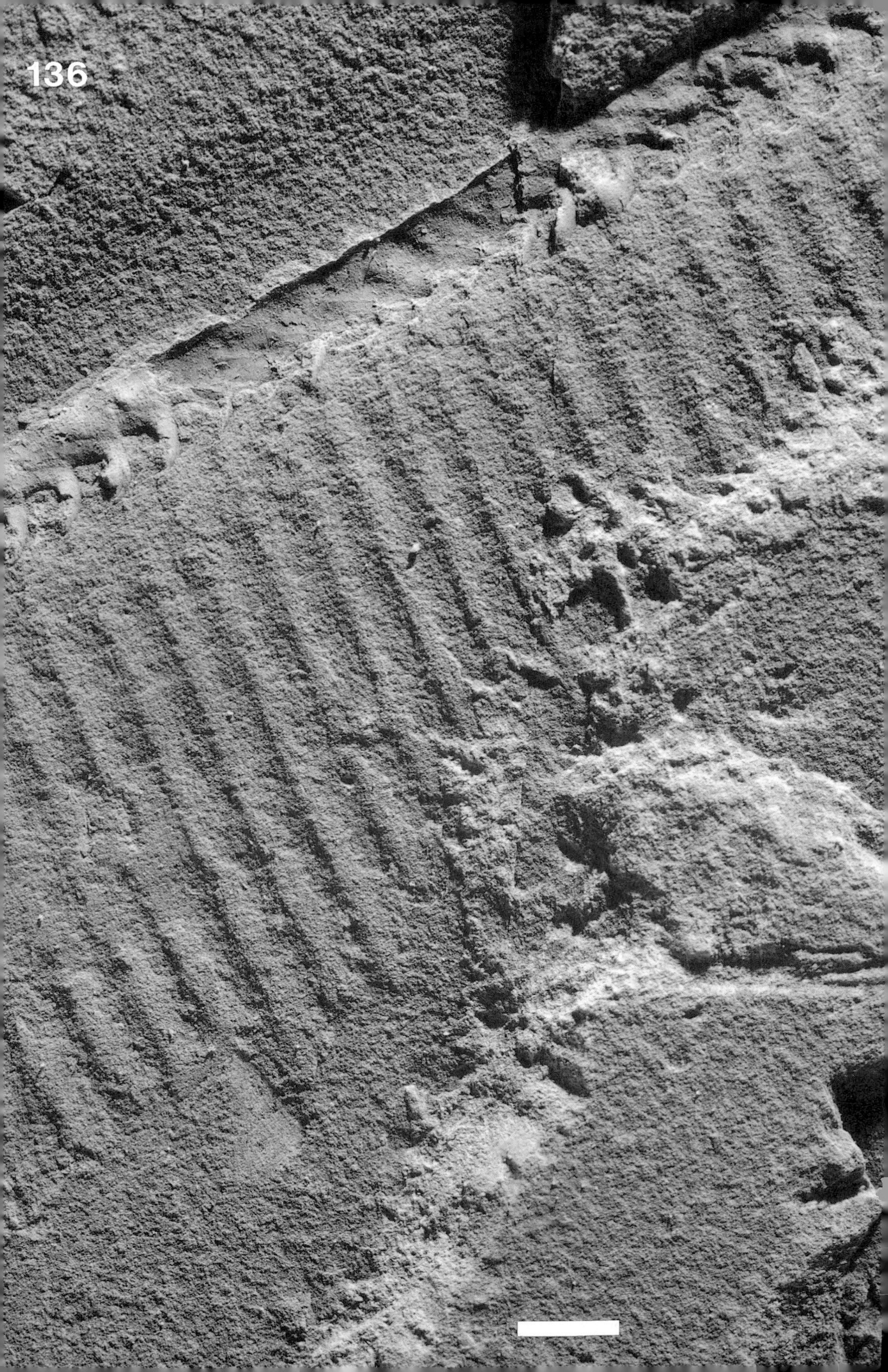
136

138

INDEX